PRAISE for BRUCE McCANDLESS II
and *WONDERS ALL AROUND*

"Bruce McCandless II waited eighteen years for his first spaceflight, marking him as an astronaut with seemingly inexhaustible supplies of patience and dedication. When he cruised serenely away from *Challenger* in 1984, photographed against the infinite cosmos, we stared in wonder at a fellow human slipping the bonds of Earth. An important and entertaining book, *Wonders All Around* gives us a complex, complete portrait of a brilliant and courageous American."

—**Tom Jones**, Veteran Astronaut and Spacewalker,
Author of *Sky Walking: An Astronaut's Memoir*

"Bruce McCandless is the astronaut the world knows from an iconic photo that has become shorthand for space exploration—and yet he's someone who few could name. Floating alone in Earth orbit, face hidden behind his spacesuit visor, his image was everywhere in the mid-80s. It was on my wall when I was a teenager, along with the usual music and movie posters. The movies and musicians came and went, yet McCandless stayed. It takes a talented author to bring old events to life, and his son's humor and candor makes me feel as close to knowing him as we could ever get. I almost feel like hanging up my old 1980s poster again."

—**Francis French**, Author of *In the Shadow of the Moon*

"When Bruce McCandless from my astronaut selection group made the first untethered space walk, photos of him became as iconic as anything taken on the moon. Bruce was a very smart guy—the brain of our group—analytical and very academic in his approach to work. So he wasn't a real 'jock,' which unfortunately for him, as things were back then, meant he didn't get to fly for almost two decades. But NASA caught up to him with the shuttle program, and his years developing that untethered EVA backpack paid off—flying it first made history."

—**Al Worden**, Apollo 15 Astronaut

"When we think of astronaut Bruce McCandless II, we think of 'the Poster': the iconic image of him free-flying in space that was a ubiquitous addition to every space geek's wall in the mid-1980s. This biography delves into McCandless's very human quirks, moods, and more unheralded achievements. His story is told alongside his family's story, and underscores what it was like to grow up and live in the shadow of NASA. *Wonders All Around* is an evocative, sometimes surreal memoir of a man and his son's childhood, and essential for astrophiles."

—**Emily Carney**, Space Historian

"*Wonders All Around*, Bruce McCandless's account of growing up the son of a real-life Buck Rogers, is at once a deeply felt memoir of fathers and sons and an engaging exploration of space and its mysteries—space beyond the bounds of Earth and space between a suburban teenager's ears. With his eyes on the stars and his feet on the ground—like his astronaut father—McCandless has produced a wonderful, readable book."

—**Joe Holley**, Author of *Hometown Texas*

"In *Wonders All Around*, Bruce McCandless III has managed to open the lid on the extraordinary life of his father, Bruce McCandless II, in a way arguably no one else could. Thanks to his fly-on-the-wall manner of writing, we meet an extraordinary man who achieved so much in his life and yet seemed to balk at showing the world how amazing he really was. We are granted a window into his inner workings and get an exposé into so much more than the man floating untethered in space. So outstanding are the accomplishments of Bruce McCandless II that we refer to his untethered spaceflight as just *another* of his amazing endeavors. Prepare to be astounded!"

—**Dwight Steven-Boniecki**, Director of *Searching for Skylab*

WONDERS
ALL AROUND

WONDERS
ALL AROUND

The INCREDIBLE TRUE STORY of
ASTRONAUT BRUCE McCANDLESS II
and the FIRST UNTETHERED
FLIGHT in SPACE

• • •

BRUCE McCANDLESS III

GREENLEAF
BOOK GROUP PRESS

This book is a memoir reflecting the author's present recollections of experiences over time. Its story and its words are the author's alone. Some details and characteristics may be changed, some events may be compressed, and some dialogue may be recreated. Furthermore, some names and identifying characteristics of persons referenced in this book, as well as identifying places, have been changed to protect the privacy of the individuals and their families.

Published by Greenleaf Book Group Press
Austin, Texas
www.gbgpress.com

Distributed by Greenleaf Book Group

For ordering information or special discounts for bulk purchases, please contact Greenleaf Book Group at PO Box 91869, Austin, TX 78709, 512.891.6100.

Design and composition by Greenleaf Book Group and Kimberly Lance
Cover design by Greenleaf Book Group, Shaun Venish, and Kimberly Lance
Cover image courtesy of NASA, photographed by Robert L. "Hoot" Gibson
Author photograph by Patrick J. Cosgrove

Publisher's Cataloging-in-Publication data is available.

Print ISBN: 978-1-62634-865-3

eBook ISBN: 978-1-62634-866-0

Part of the Tree Neutral® program, which offsets the number of trees consumed in the production and printing of this book by taking proactive steps, such as planting trees in direct proportion to the number of trees used: www.treeneutral.com

Printed in the United States of America on acid-free paper

20 21 22 23 24 25   10 9 8 7 6 5 4 3 2 1

First Edition

For Mom, Dad, and Tracy.
Are we there yet?

Those who dwell, as scientists or laymen,
among the beauties and mysteries of the earth
are never alone or weary of life.

—Rachel Carson, *The Sense of Wonder*

Fella, you don't start playin' ball at your age. You *retire*.

—Pop Fisher to Roy Hobbs, *The Natural* (1984)

"Bruce McCandless and I were both members of the Class of 1958 at the United States Naval Academy. As an undistinguished graduate of that class, I always looked up to Bruce—not only for his incredible intellect, but also for his character and integrity, which embodied the highest values of the United States Navy. The iconic photo of Bruce soaring effortlessly in space has inspired generations of Americans to believe that there is no limit to human potential."

—SENATOR JOHN S. MCCAIN

CONTENTS

FOREWORD

I grew up in the same fantastic time and place as Bruce McCandless III—in the shadow of the Johnson Space Center in the late sixties and seventies. Bruce and I weren't close childhood friends, but our paths crossed a few times. Because we lived in Texas, where all boys are required by law to play football, we proudly donned the blue and white of the Pop Warner league Comets. Bruce was a cornerback. I was a defensive end. We laced up our pads, fought for our turf, and even beat our arch-rivals, the Space Bandits. We went on to attend the same intermediate and high school, and, like many in our high school class, the same university . . . Hook 'em Horns! But it was really only after connecting on social media as adults that we became friends. Knowing my interest in space exploration based on some of my online posts, Bruce contacted me and said he was working on a book about his dad and NASA. When he asked if I would be interested in reading a draft and providing comments, I jumped at the opportunity. I think Bruce was looking for a constructive review of the technical details related to the NASA missions—an appropriate expectation, given that I went on from the University of Texas to an information technology career in the aerospace industry and he seems to have spent a lot of time writing sonnets.

But for those of us who grew up around the space program, it's hard to remember those days in anything other than personal terms—families and friends, crushing losses and shared celebrations. I ended up providing more of an anecdotal recollection of my childhood growing up in Timber Cove through the Mercury, Gemini, Apollo, Skylab, and early space shuttle missions. My father, Cecil Kelly, was one of the over 400,000 men and women who worked with NASA in the sixties to achieve America's goal of putting a man on the moon and returning him safely before the end of the decade. Goal Accomplished!

Like a lot of us who grew up during the Apollo years, I didn't really appreciate what I was witnessing until much later in life. I look back on those days now with amazement. As I reviewed this well-told story, I would share with Bruce some related tale (like one about the poem my mom, Sharon, wrote to our neighbor Jim Lovell after Apollo 8, or a remembrance of our friend Jack Kinzler, a completely normal-looking NASA engineer down the street who contributed strange but brilliant ideas in support of the Apollo and Skylab missions). A few weeks later, Bruce sent me something in the mail. I laughed out loud when I found an old-school Apollo 10 Snoopy sticker inside. I told Bruce how much I enjoyed reliving my childhood memories as I learned more about the McCandless family, warts, injured owls, and all. I'm sure the process of writing this book was an emotional roller coaster, and I appreciate him for seeing it through.

In all honesty, Bruce and I didn't really bring glory to the gridirons of Saturday morning football in Texas. No matter. We found different dreams to chase. Bruce became a gifted writer, with the ability to make you feel like you are right there with him as he relives the stories, memories, and emotions surrounding the Incredible True Story of Bruce McCandless II. For me, the book was a re-entry. For you, it may be a launch. But if you're interested in space, and you like learning about the people who got us off Earth and into the heavens, you're going to love *Wonders All Around*.

Chris Kelly
Plano, Texas
June 2021

ACKNOWLEDGMENTS

First things first: This book wouldn't exist without the support, enthusiasm, and occasional arm-twisting of my wife and co-conspirator, Pati Fuller McCandless. Roger G. Worthington, Grant Tandy, and Kody Osborne of the Worthy Garden Club started me on this project and made me follow through even when they were busy saving the world. Claire Bowman, Fred Collins, Jim Crocker, June Fuller, Gayle Gordon, Chris Kelly, Ellen Shields McCandless, Chris Parks, Mark Robbins, Martha Robbins, Jeanette Scott, Don Sembera, Ron Sheffield, and Kathy Sullivan kindly read, criticized, corrected, and greatly improved this memoir. Walter W. "Bill" Bollendonk, Dr. Ben Clark, Robert L. "Hoot" Gibson, Fred Haise, Jim Hornfischer, Jack Keeney, Dr. Joseph P. Kerwin, Jack Lousma, David Paul, David G. Welch, and Susy Young provided important information included in the story. Expert editors Sally Garland, Joan Tapper, and Pam Nordberg brought clarity and cohesion to my manuscript. Shaun Venish helped me figure out what the book should look like, and Kimberly Lance at Greenleaf made it happen. My sister Tracy McCandless remembered things I didn't and imagined things I couldn't. Filmmaker Dwight Steven-Boniecki shared his vast knowledge of Skylab matters with me. Artist Ed Hengeveld provided valuable assistance in finding and identifying many of the photographs contained in these pages. The elegant author Francis French asked great questions and then helped me find the answers. Veteran astronaut Tom Jones knows a lot about everything and provided firsthand knowledge to correct my erroneous assumptions about space travel. The extraordinary Emily Carney encouraged me every step of the way, and the Space Hipsters Facebook group was a constant source of inspiration, amazement, and humor. Thank you—all of you—for your help. Any mistakes or misstatements contained in *Wonders All Around* are my fault, not yours.

Note: Some of the material contained in this book appeared previously as "From Arcturus to Armand Bayou" in the 49th episode of the podcast *A Minute of Your Time*, released on May 27, 2020.

A WORD ABOUT NAMES

In my father's family, names repeat themselves. I have four female ancestors named Velma. My great-grandmother, my grandmother, my dad's sister, and her daughter were all named Sue. In the same way, my grandfather, my father, and I are all named Bruce—Bruce I, Bruce II, and Bruce III. My grandfather and I both figure into this biography, but mostly at the edges. Thus, while I identify my father as Bruce McCandless II on several occasions early in the narrative, I mostly just use the name Bruce McCandless as the story goes on. It's easier that way, and I suspect you'll know who I'm talking about.

NOTABLE DATES

October 12, 1919   WWI veteran Anthony "Buck" Rogers rendered unconscious by toxic gas in Pennsylvania mine; falls asleep for 500 years.

June 8, 1937   Bruce McCandless II born in Boston.

September 1946   Astronomer Lyman Spitzer discusses advantages of space telescope in his paper "Astronomical Advantages of an Extra-Terrestrial Observatory."

June 4, 1958   Bruce McCandless II graduates from U.S. Naval Academy.

August 6, 1960   Bruce McCandless II marries Alfreda Bernice Doyle.

May 5, 1961   Alan Shepard becomes first rocket-borne American in space.

May 25, 1961   President John F. Kennedy challenges the United States to land a man on the moon and bring him safely back again by end of the decade.

August 15, 1961   Bruce McCandless III born in Annapolis, Maryland.

February 20, 1962   John Glenn is first American to orbit Earth, on third manned Mercury flight.

July 13, 1963   Tracy McCandless born in Virginia Beach, Virginia.

June 3, 1965   Gemini 4 astronaut Ed White becomes first American to walk in space; uses handheld maneuvering unit.

April 4, 1966   Bruce McCandless II selected as youngest member of astronaut Group 5.

June 5, 1966   Gene Cernan fails in attempt to test Astronaut Maneuvering Unit during Gemini 9 "space walk from hell."

January 27, 1967   Fire in Apollo 1 capsule kills Gus Grissom, Ed White, and Roger Chaffee.

July 20, 1969   Neil Armstrong and Buzz Aldrin walk on the moon; Bruce McCandless II is capsule communicator (capcom).

1973–74 Bruce McCandless II trains as backup pilot for first crewed Skylab mission; predecessor to Manned Maneuvering Unit (MMU) tested inside orbital workshop by astronauts on second and third crewed missions.

April 12, 1981 First orbital flight of the space shuttle.

February 7, 1984 Bruce McCandless II and Bob Stewart fly the MMU on tenth shuttle mission, designated STS-41B.

January 28, 1986 Shuttle *Challenger* destroyed shortly after launch, killing six astronauts and schoolteacher Christa McAuliffe.

February 1 and 5, 1990 Cosmonauts test Soviet version of MMU while tethered to the *Mir* space station.

April 25, 1990 STS-31, with mission specialists Bruce McCandless II, Kathy Sullivan, and Steve Hawley, deploys Hubble Space Telescope.

August 31, 1990 Bruce McCandless II retires from NASA, joins Space Telescope Science Institute panel to study possible fixes for Hubble's faulty main mirror.

December 2–13, 1993 Shuttle mission STS-61 repairs Hubble Space Telescope.

September 9, 1994 STS-64 astronauts test Simplified Aid for EVA Rescue (SAFER) device, fashioned from MMU technology.

November 2, 2000 First crew (2 Russians, 1 American) takes up residence in International Space Station.

February 1, 2003 Shuttle *Columbia* destroyed during reentry over western United States, killing all seven astronauts aboard.

July 8, 2011 STS-135 is final flight of the space shuttle.

January 14, 2014 Bernice McCandless dies in Conifer, Colorado.

December 30, 2014 Bruce McCandless II marries Ellen Shields in Denver.

December 21, 2017 Bruce McCandless II dies in Los Angeles.

April 17, 2423 Buck Rogers emerges from cave to behold strange and terrifying future.

1

THE LONG DRIVE

He was an Apollo-era astronaut, one of nineteen men picked from hundreds of qualified applicants to plant the American flag on the moon. Now, though, the Apollo program was over, and Bruce McCandless's prospects for leaving Earth seemed bleak.

Some men would have started looking elsewhere for a job. My dad's reaction was just the opposite. He kept pushing for a spot on some future spaceflight, even if the flight was, at this point, largely hypothetical. He worked hard. He checked all the boxes. And when Congress enacted a nationwide 55 mph speed limit in an attempt to reduce oil consumption, my father dutifully complied. *Dad! A fighter pilot!* A man who'd wrestled a Phantom warplane capable of flying 1,200 miles per hour onto the deck of a lurching aircraft carrier in a thunderstorm, at night, was now poking along Highway 183 north of Austin in a barn-size Chevy Suburban with the speedometer pegged on double nickels. Even worse, I was a fifteen-year-old boy stuck in this clown car with him and the rest of my family.

My father manned the driver's seat, silent except when he saw some natural or historic feature in the distance. "The Llano River!" he announced, as if he'd put the waterway there himself. My mother, Empress of GORP, nodded companionably from the passenger side, content to indulge her husband's periodic travelogue but focused mostly on smuggling snacks to us in the back and watching the wispy winter clouds. My little sister, Tracy, the Precious Cargo, paged through *Tiger Beat* in the middle seat, and I occupied the way-back—the Tail Gunner, as I thought of myself, strafing the countryside with my piercing scowl.

It was January 1976. West of Abilene, the empty sorghum and cotton fields of the Panhandle lay around us like a sea of dirt. Bulbous water towers were the only landmarks, guarding the horizon like eyeless pod creatures on spindly metal legs. I watched plumbing vans accelerate past our car. Eighteen-wheelers ribboned like long insects from our lane to the next and then back again once they were safely in front of us. Cadillacs full of languorous suburban kids from Dallas swung wide as they went by, the passengers barely sparing a glance at us as they sped on their way to Taos or Aspen. We passed through dreaming skeleton towns: Sweetwater, with its sign for the world's biggest rattlesnake round-up; Snyder, where a long-ago resident had spotted a beautiful and extremely rare white buffalo, and then killed it; and Post, the cereal millionaire's not entirely successful agrarian utopia.

It took me years to figure out why a man with the skills and experience to drive as fast as his V-8 could go would possibly submit to the idiotic strictures of a speed limit *no one else* in the great state of Texas was paying the slightest attention to. But I get it now. Such were the vagaries of NASA's flight-crew selection process that anything—marital scandal, a five-car pile-up, a negative fitness report—might have gotten Bruce McCandless scratched from the lineup for a future spaceflight. And he was not going to let that happen. So he toed the line, and we toed it with him.

Unfortunately, Dad's insistence on abiding by the rules seemed to make little difference. As early as 1973 he was being called a failure by the press—one of only three men from his astronaut class, Group 5, not to have been selected for or actually flown on either an Apollo or a Skylab mission. Indeed, excluding Ed Givens, who died in 1967 from injuries sustained in a car crash, and John Bull, who left the program in 1968 for medical reasons, and including projected Apollo, Skylab, and Skylab rescue flights, he was the *only* man not to be picked. A journalist interviewed my father as he worked as a capcom—capsule communicator—for Skylab 2, and the story showed up in newspapers all around the country. The writer called Bruce McCandless a "forgotten astronaut" and concluded that "declining budgets, national priorities in flux and the cruelty of time lengthen the odds [McCandless] ever will exult in the thunder and fire of launch, float weightless or wear the gold astronaut pin that separates the 'been theres' from the 'somedays.'" It was a humiliating experience for my father. He was a brilliant man. He had always been something of a wunderkind, marveled at for his mathematical aptitude, capacious memory, and rock-ribbed self-discipline. Now the prodigy, son and grandson of naval heroes, was being portrayed as a washout in the national media. He was the Moonlight Graham of the space program, the promising rookie ballplayer who never even got to the plate in the big leagues.

There was a mission the journalist failed to mention. Still in the planning stages in 1973, when the article was written, the Apollo-Soyuz Test Project was a one-off American-Soviet flight meant to mark the end of the space race and usher in an era of friendlier relations between the United States and the USSR. America's astronauts would leave Earth using Apollo hardware. The Soviets would fly their Soyuz craft, and the two vehicles, Soyuz and the Apollo command module, would link up in orbit using a specially designed collar that would allow the two crews to go back and forth between the spaceships. Dad prepared a cheat sheet to pitch himself to management as a candidate for the mission. He noted that he'd studied German at the Naval Academy and was currently teaching himself Russian, that one of his hobbies was photography, and that when he served as a young lieutenant on the USS *Enterprise* he'd acted as an informal ambassador for foreign dignitaries visiting the ship. He admitted he was occasionally "off-putting"—*too damn smart* is what he meant, *and not afraid to show it*—but insisted he could work with anyone in the astronaut office.

Alas, he wasn't chosen for that mission either. The three slots were allotted to Tom Stafford, Vance Brand, and the grizzled veteran Deke Slayton, finally cleared for spaceflight after a long medical suspension for heart issues. The mission went well. Linked together, the space vessels looked like two bugs kissing through a harmonica. Our astronauts shook hands with their cosmonauts, Alexei Leonov and Valeri Kubasov. The combined crew made toasts and constructed a commemorative plaque. It was a nothing flight, a symbolic gesture, but it was a *flight*, and my dad's frustration deepened when he wasn't picked for even this relatively undemanding assignment.

I remember my father's days of waiting for a mission as a sort of perpetual twilight. Bruce McCandless didn't smoke or drink. He dutifully jogged several miles each week on the oak-shaded track behind the astronaut gym. He avoided TV during the week and watched only sparingly on weekends, when my mom could coax him into a spree of slothfulness that included *The Mary Tyler Moore Show* on Saturday evenings and *Masterpiece Theatre* on Sundays. He didn't watch sports, play sports, or gamble on sports. He didn't fish or hunt, and he never cracked open a cold one with his buddies. In fact, I don't remember him having any buddies. He was a loner, courteous but self-contained. You didn't get a pat on the back in our house unless you were choking.

Our dinners were somber affairs. We ate around a rectangular Formica table in the breakfast nook. Tracy and I sat on benches padded with orange vinyl cushions. Mom and Dad occupied *faux*-Spanish style chairs with green felt upholstery. Despite the informal, Howard Johnson's-at-the-airport feel of the furnishings, there was a tension in the air that set in right around the time the frozen string beans started steaming. I had the feeling that my sister and I had forgotten to do something important, though I couldn't figure out what it was, or that judgment had been rendered on us and we'd been found guilty of . . . *something*—again, it was unclear what. Horseplay was prohibited. The TV and all sources of music or other frivolity were turned off, and singing was strictly forbidden. The only sound came from the aquarium pump. My father had a 100-gallon tank along the wall behind his chair. Sometimes the big plecostomus would attach itself by its mouth to the glass facing us, and I imagined it was sucking all the oxygen out of the room.

But one night it was worse. One night Dad, a resolutely even-keeled man, unflappable really, slumped in his chair, put his head in his hands, and went completely silent. It was such an odd and disturbing sight that gravity itself seemed to bend and fold around his motionless figure. We'd seen him mad plenty of times. He hated the length of my hair. He was annoyed when I said I wanted to be a DJ when I grew up. He got mad at my mom and frustrated with Tracy, just as we all—like any family—got angry and annoyed with each other. But this was different. This was an existential crisis in the life of a man who didn't have existential crises.

The room was quiet, save for the rasp of the pump. It was like the house itself was on life support. Was my father mourning his lost potential? Was he bitter about the years he'd wasted, waiting for a flight that never came? Was he physically ill? All three? Tracy and I had probably been bickering. And maybe there was more to it. A disagreement with my mother. The pressures of getting by on a government paycheck. But I suspect the biggest frustration for my dad was the feeling that he was stuck. He'd always been at the top of his class. He'd succeeded at everything he tried. Now he found himself a man without a mission, branded as a failure in the media, an astronaut who would never see the stars. The Skylab program had been over for two years, and the first flight of the space shuttle was still half a decade in the future. There were rumors going around about selection of a gigantic new astronaut class, a host of young hotshots just as eager to see space as Bruce McCandless was. He knew he could pass whatever test he was assigned. It was just that no one would give it to him. So, for the moment, he sat. And the house was very quiet, the way a forest grows quiet before a storm.

The next month our family packed up for the long drive to Denver, where Dad was training at a Martin Marietta aerospace facility and where my mom, my sister, and I were going to tour the U.S. Mint and visit the Denver Zoo. We brought extra socks and sprayed our jeans with Scotchgard because there was a chance we'd get to go skiing. The trip from Houston took twenty-three hours. Every time a car zipped past us, I'd check Dad's face in the rearview. There wasn't a trace of emotion on it. But periodically he would lean forward in his seat and crane his neck to gaze out the windshield to where jet aircraft, possibly fighters out of Dyess Air Force Base, were painting bars across the sky with their ghostly contrails.

A decade had passed since my father left his PhD studies at Stanford University to join NASA, lured by the prospect of a walk on the moon and a place in the history books. What had once seemed like a sure bet was now fading on the far turn. He knew he might never get beyond the blue veil to the vast unknown beyond. What he didn't know was why. We didn't know either. Was it something about him? Was it us? Was it, I wondered, *me*? At fifteen I had discovered I was a fundamentally flawed and repugnant human being, full of violent and generally unprintable desires. It was entirely possible I was the Jonah on this damned ship, holding my father back from what he'd wanted for so long. True, he never said as much. The fact was, he never said much at all. But we felt his disappointment. We breathed his frustration. And even in the way-back of the Suburban I could see that his knuckles were white as we left Lubbock for Amarillo, heading north through the brown lands toward the peaks and forests of Colorado. I wasn't sure we'd make it. Jesus. *The miles.* The speedometer remained at 55, and it was obvious.

We were doomed.

We were still driving six hours later.

"Raton Pass!" shouted Dad, and we started to climb.

2

BATHED IN BLOOD

Bruce McCandless II was born in Boston in 1937, heir to a peculiar kind of fortune. In an age when social ties were still important in the United States Navy, the most clannish and tradition-bound of the services, he was the son and grandson of decorated officers. His surname was a set of epaulets, a commendation in itself. And while family ties might have opened doors for him if he'd asked, he didn't need any help to make his own mark on the world. Though you may not recognize his name, you've seen Bruce McCandless's picture in books and on magazine covers, in TV commercials and spaceflight documentaries. As planetary scientist Dr. Ben Clark puts it, "He's the one astronaut everybody knows, whether they know it or not."

The family line is bathed in blood. My dad's great-grandfather, David McCanles, was ambushed and killed in July 1861 by a ginger-haired stable hand named James Butler Hickok. Tales of the murder, much embroidered by the killer's sycophants, gave the West one of its seminal legends, and the redheaded gunman, later known as Wild Bill Hickok, went on to become a storied assassin and enforcer, a messenger of death in a dandy's waistcoat. Stories followed him wherever he went. Whether he was a hero or a villain differed from one man's telling to the next, but there was no disputing that Hickok was a gifted killer, that he walked with dark angels and brought bad luck wherever he went. He was finally dispatched by an angry gambler who took him from behind, shooting him through the head as the famous gunslinger stared at his cards: black aces and eights, what we now call the Dead Man's Hand.

David's son Julius moved west, eventually settling in Florence, Colorado. Like other members of the family, he changed his name from McCanles to McCandless to avoid association with Hickok's slaying of his father. Julius's son, my paternal great-grandfather Byron McCandless, left Colorado in 1901 to attend the United States Naval Academy. He went on to become an inventor, a flag historian, and a naval officer of some note who was called out of retirement to command America's most important destroyer base during World War II. He was known for his tendency to tour the facilities where workers welded, riveted, and sweated steel into seaworthy weapons. As rugged as a railroad spike, he was partial to grabbing a hammer and joining in the labor—a predilection that earned him the nickname "Captain Bing Bang" from his men. Byron supposedly taught my dad the rudiments of calculus when my father was only seven, and the old man's flinty, do-it-yourself attitude left a lifelong impression on the boy.

Dad's maternal grandfather, Captain Willis W. Bradley, Jr., was equally accomplished. A Congressional Medal of Honor winner who was appointed governor of Guam in 1923, he fought the powers that be to improve the political lot of Guam's native Chamorro people and eventually retired to California, where he served as a congressman and, later, a state assemblyman. My dad enjoyed recounting the accomplishments of his grandfathers. Still, it was *his* father—undersized, underestimated, the unlikely survivor of a naval nightmare—who had the biggest influence on Bruce II. Any story about my dad has to start with the first Bruce and his miraculous escape from a cauldron of fire.

MY FATHER WAS FOUR years old and living with his family—dad, mother, and little sister—in a rental house on Oahu when the Japanese attacked Pearl Harbor on December 7, 1941. Lt. Commander Bruce

McCandless was a junior officer serving on the heavy cruiser USS *San Francisco*, which was stationed at Pearl. It was early on a Sunday morning when the Empire of Japan's initial wave of 180 aircraft appeared from the north, their propellers whining like angry bees. My grandfather was asleep when the first bombs fell. His wife, Sue, woke him up by telling him about the smoke she could see from the kitchen.

"If it's white," he said, "it's nothing to worry about."

"Nope," she replied, swallowing her fear. "Pitch black."

The long-anticipated conflict had finally begun. As the attack continued, young Bruce wandered outside to watch dark plumes of smoke from burning fuel oil rise from the crippled ships in the harbor below. His father dressed hurriedly and drove down to Pearl. He was stopped at the gate by Marine guards fearful of sabotage, who, in true jarhead fashion, ignored all appeals to logic and made the young Navy officer get out of his car and proceed on foot. Bruce started running. He made it to the *San Francisco* just in time to see another wave of Nakajima B5N "Kate" bombers come roaring in over the harbor. Bruce strapped on a steel helmet, grabbed his .45 caliber pistol, and started pumping the skies full of lead. When the attack was over, he stayed at Pearl to join repair and rescue operations. He stayed all day and most of the next. In fact, aside from a quick visit home by Lieutenant Commander McCandless to reassure his family, my dad didn't see his father for almost a year.

The attack was a disaster for the Navy. The Japanese killed 2,300 Americans, mostly naval officers and sailors, and injured another 1,100. Nineteen U.S. ships were sunk or damaged, and almost 200 airplanes destroyed. My grandmother, my dad, and his little sister were evacuated from Oahu not long afterward, amid fears that the Japanese would invade Hawaii. On Christmas Day, Sue and her children set off across the Pacific aboard the *Lurline*, pride of the Matson passenger fleet but stripped now of her finery—a living room for Yankee wives and frightened tots. Through winter seas they steamed toward the West Coast, haunted by rumor and a thousand spectral sightings: the outlines of aircraft in an oyster gray sky, the stalk of a submarine scope to starboard, glimpsed for a moment through wind-tossed waves. Sue carried flatware, life-insurance policies, and a medical kit she'd bought for young Bruce, his only Yuletide gift. The three McCandlesses ate powdered eggs off paper plates and lived in fear of the next bad news. Some passengers traveled with more than memories. The remains of their menfolk lay in pine boxes in the cargo hold. Sue wrote letters on the backs of envelopes, wondering if she could hold her children if the ship went down or, worse, could hold only one of them. She wondered where her husband was and dreaded what the future had in store for the people around her, stunned by all they'd seen. Back on Oahu, my grandfather was granted a brief period of liberty just after Christmas. He drove home to find the rental house empty, his wife and children already at sea.

My dad never forgot Pearl Harbor and his evacuation from Hawaii. He grew up feeling as if he and his family, rather than the country generally, had been attacked on that long-ago Sunday morning, and there was never any doubt in his mind that he would follow his father and grandfathers into the service. He was already, at the age of four, a Navy man.

My grandfather Bruce was an unlikely hero, a soft-spoken individual who loved history and literature, gentle satire and clever puns. He served as communications officer aboard the *San Francisco* during the First Naval Battle of Guadalcanal, a nightmarish fight in November 1942 in which the American cruiser faced off against two Japanese battleships, the 37,000-ton *Hiei* and her sister ship *Kirishima*. They were the Death Stars of their day, heavily armored and bristling with 14-inch guns that could land 1,400-pound shells on targets more than twenty miles away. Navy intelligence reports indicated that *Hiei*, *Kirishima*, and a number of other Japanese warships had been sent to bombard Henderson Field, the U.S. airstrip on Guadalcanal. Choosing the *San Francisco* (or "*Frisco*," as the crew sometimes called the vessel) for his flagship, Admiral Daniel Callaghan set out on November 12 with four other cruisers and eight destroyers to intercept the attack. His confrontation of the Japanese that night was intentional but not well planned. Perhaps no planning was possible. He was facing a vastly superior force, with no clear orders but to engage the enemy. It was a suicide mission, and every officer knew it. The sailors kept their own grim counsel. As the clock passed midnight, they noted it was now Friday the thirteenth and that their battle group, Task Force 67, numbered thirteen ships.

The Japanese advanced from the northwest under cover of a violent squall. They emerged from the rain into a moonless night, the sky obscured by low clouds, the fragrance of gardenias from nearby islands drifting over the sea. The opposing forces failed to spot each other until they were disastrously close. Outnumbered and outgunned, the Americans had one unexpected advantage: The Japanese had prepared for their bombardment of Henderson Field by loading their guns with incendiary and high-explosive shells rather than the armor-piercing variety. While incendiary shells were cruelly effective in killing men, they were less than ideal for battle with a steel-plated opponent.

"Get the big ones first!" Callaghan ordered when he saw the Japanese ships. "We want the big ones first!" What followed was a reasonable facsimile of hell, a confused and vicious melee of a sort that will probably never be seen again. The American destroyer *Laffey* passed within twenty feet of the *Hiei*—close enough that the crews of the respective vessels could have battled each other with spears, had any been handy. At such close quarters, the Japanese ship seemed to one American officer to be "the biggest man-made object ever created." Searchlights stabbed the darkness. Starshell flares drifted seaward from above, while tracers arced blue, red, and white through the humid air. The gray guns thundered. Ships appeared and then vanished again in the pulse of shellfire. Men screamed and fell in the storm of steel, cut in two by sprays of shrapnel or blown overboard by the blasts.

Severely outclassed, the *San Francisco* nevertheless opened up on the *Hiei* with her eight- and five-inch guns at a range of just over 2,000 yards, scoring hits that crippled the monstrous vessel's steering engine and contributed to a massive fire that soon engulfed the battleship's superstructure. The *Frisco* was rewarded for her effrontery by fire from not only the enraged crew of the *Hiei* but also from the *Kirishima*. In the following few minutes, the cruiser was shredded by forty-five Japanese shells and at one point had twenty-five fires burning on board. A series of Japanese shell strikes tore through the *Frisco*'s bridge, killing almost every officer present, including Admiral Callaghan and Captain Cassin Young. Stunned but still in possession of his faculties, Bruce McCandless pulled himself up off the steel floor to realize he was now the senior officer on deck. Torn and bleeding from a shrapnel wound, severely concussed, the young officer nevertheless took command of the cruiser, attempted to reestablish the American battle line, and brought the ship back into the fight. Despite heavy damage and the fact that she was taking on dangerous amounts of water, the *Frisco* peppered an approaching Japanese destroyer with fire from her port-side five-inch battery. More significant was the damage the *San Francisco* had done earlier to the *Hiei*'s steering ability. Tracked down and bombed by American airplanes, the crippled *Hiei* was sunk later that day. Japanese admiral Isoroku Yamamoto was reportedly so infuriated by the loss of his prized warship that he removed her commander from duty and had him forced out of the Imperial Navy.

Despite its ferocity, the battle was inconclusive. While the Japanese withdrew from the fray without accomplishing their objective, they inflicted serious damage on Task Force 67. The *Laffey* and several other American vessels were sunk as a result of the engagement, and some 1,400 officers and enlisted men lost. All five of Iowa's fighting Sullivan brothers went down with their ship, the USS *Juneau*. Still, Henderson Field survived another night, and after a series of similarly brutal encounters between the two navies, the war in the Pacific eventually turned in America's favor. It was, after all, a conflict of wills as well as weapons. The Imperial Navy always knew its enemy could fight; what seems to have come as a surprise is that Americans *would* fight, and with a valor and viciousness at least equal to that of the Emperor's men. This was the significance of Guadalcanal.

For his actions on November 13, 1942, Bruce McCandless was awarded the Congressional Medal of Honor, the nation's highest military decoration, and promoted to the rank of commander by order of President Franklin D. Roosevelt. Roosevelt called the engagement "one of the great battles of our history." It was a time when the nation was searching for silver linings in the clouds of war. Newspapers and magazines across the country carried the story of the "stocky, good-looking" McCandless and the *Frisco's* unlikely triumph against long odds. "Nothing in the long and hallowed history of our naval forces," opined one journal, "can exceed the amazing feat accomplished by young McCandless." The *Long Beach Press-Telegram* ran a front-page photograph of five-year-old Bruce II playing with a toy boat, with the caption: HIS DAD'S REAL HERO.

AFTER REST AND REHABILITATION back in the States, Commander McCandless returned to sea. He was eventually given command of the destroyer USS *Gregory*, which, in April 1945, participated in the occupation of Okinawa. As Marines stormed the island, the last stepping-stone before a planned invasion of the Japanese home islands, the *Gregory* was attacked by three suicide planes—the dreaded kamikazes that served as human sacrifices for the rapidly fading glory of imperial Japan. Though most kamikazes were brought down before they could damage their targets, a few inevitably slipped through the screens of anti-aircraft fire. They managed to sink thirty-four Allied ships during the war and damaged numerous others. During the Okinawa campaign alone, they were responsible for the deaths of an estimated 5,000 American servicemen. Though two of the planes that attacked the *Gregory* that day were shot out of the sky, a third struck the vessel amidships on the port side, damaging her forward engines. The *Gregory* was knocked out of service by the attack. The ship eventually limped back to the West Coast, where it berthed for repairs at the U.S. Repair Base, San Diego, under the command of Commodore Byron McCandless. A headline captured the situation: CMDR. MCCANDLESS TAKES SHIP TO FATHER'S YARD FOR REPAIRS.

"They sort of folded your ship up like an accordion, son," Byron remarked of the damaged destroyer, a mass of twisted steel and baked-on blood. Captain Bing Bang was already sizing up the vessel for repairs. "Can you play a tune on that thing?"

3

THE BOY
WITH TWO BRAINS

Despite the fact that the world was becoming an increasingly dangerous place in the 1930s, life in the United States was largely peaceful, if not prosperous. Still struggling to shake off the economic effects of the Great Depression, the country suffered through another setback in 1937 when unemployment levels, previously falling, spiked again, rising to 19 percent over the course of the following year, and industrial production levels declined precipitously. My grandfather McCandless's commission as a naval officer wasn't lucrative, but it provided a steady income. Because officers' wives rarely worked outside the home in those days, Sue Bradley McCandless had little money but plenty of time to mark family milestones like the birth of a first child. Indeed, it would be hard to imagine a more lovingly chronicled infancy than my dad's.

Sue was a stylish and briskly intelligent woman with the same arched eyebrows and delicate chin as the movie star Myrna Loy. The new mother kept notes on her son's every developmental phase. Thus, an olive-drab tome called *Our Baby*, referred to in the family as the "Baby Book," tells us that Bruce McCandless II was born at 2:36 a.m. on June 8 (a Tuesday), 1937, at Boston's Massachusetts General Hospital. (He was actually born with a different name, *Byron Willis* McCandless, in honor of his two grandfathers, but Sue had it changed by court order on June 6, 1938. It's unclear why.) The attending physician was Dr. Benjamin Tenney, Jr.—no word on his hair color or educational pedigree—and my dad's hospital number was Baker Memorial 2497. He was twenty-one and one-half inches long at birth and weighed seven pounds, six ounces. He smiled for the first time on June 29, 1937. His first water bath took place on July 8, and he got his first shoes—Mrs. Day's Ideal Shoes, it turns out, size 3, with tiny pink roses on the instep—on February 14, 1938.

Young Bruce, who would eventually earn a master's degree in electrical engineering and get within a few ohms of a doctorate, displayed early what were to be lifelong inclinations and interests. He learned to crawl at eight months but would only go backward. On the Things Baby Likes Best page of the Baby Book, his mother listed "lights—plugs—cords—toasters—radios—anything electrical or mechanical!" At the age of one, the boy placed a pile of calling cards in the toaster, occasioning a small fire, and happily gnawed on the cords of lamps and the family radio. A set of rudimentary schematics he drew at the age of two and a half depicts a "toaster, toast, plug & cord," a "light bulb and plug," an electric razor, and a "clock & cord & plug." He received his first haircut in January 1938 aboard the destroyer tender USS *Rigel*, an *Altair* class vessel named for the brightest star in the constellation Orion. His favorite nursery school teacher was a woman named Mrs. Eureka Forbes.

Sue McCandless kept meticulous notes on the first several years of Bruce II's life, but the notes grow sparse after 1945. I know he spent some of that time in South Carolina. His father was stationed there after the war, commanding a squadron of mine-laying vessels based at Charleston, and Dad was occasionally able to join him during maneuvers at sea. My dad was oddly tight-lipped about his upbringing. He told me in 2016, shortly after I'd visited North Carolina with my family, that he'd attended a summer camp near Asheville as a kid. It was beautiful, he said. He loved it. It was the first time in fifty-five years I'd heard a word about it. (Sure enough, the Baby Book confirms that he spent eight weeks at Camp Sequoia, near Asheville, during the summer of 1950.) I know he attended the exclusive Severn School in Annapolis from 1950 to 1952 and that he was an avid reader.

He started talking about going into space at the age of three, according to his mother. As he grew older, he loved reading the adventures of space ranger Buck Rogers in the Sunday comics—an influence that would resurface some years later to play a major role in his professional life. He devoured the works of Holling Clancy Holling and C. S. Forester but gravitated eventually to the popular science publications of Willy Ley, especially *The Conquest of Space*. Ley writes with clarity and enthusiasm about Keplerian ellipses and atmospheric escape velocities. Still, it's the illustrations of artist Chesley Bonestell, a sort of Frank Frazetta for the oscilloscope set, that occupy the imagination. Tiny human figures spill out of needlelike rockets on shadow-haunted asteroids. Saturn fades on some far-flung lunar horizon, and the Beta Lyrae binary star looms in a sky swirling with vermilion menace. It's hard to say what impressed twelve-year-old Bruce more: the uncanny illustrations or the orbital arcana, the formulas or the fantasies. The budding engineer pored over the pages for hours, imagining himself to be one of those intrepid space explorers in the purple dreamscapes, drilling away on Jupiter or prospecting for signs of life on some odd and lonely megamoon. In a publicity photograph taken of him just before he headed to the United States Naval Academy, Bruce sits with his feet up on his grandfather's desk, surrounded by framed photographs and maritime tomes. He's reading a book, and he's careful to make sure we can see the cover. It's another work by Willy Ley. This one is called *Rockets, Missiles, and Space Travel*.

"I want to walk on the moon," he'd tell his parents. Also his siblings. In fact, anyone who would listen.

"That's nice," his father would respond. "But don't get your hopes up. Men won't be going into space anytime soon."

I HAVE ONLY A vague sense of where my dad was and what he did during those years. He may have been unhappy. It was around this time that my grandfather Bruce began to experience tingling and numbness on the right side of his body. He was treated for these symptoms at Bethesda Naval Hospital and eventually felt well enough to return to duty, this time in a variety of administrative jobs at the Naval Academy. The respite was short-lived. During a subsequent stay at Bethesda, my grandfather was told that his future held progressive paralysis. No one knew exactly why it was happening, though there was a vague diagnosis of multiple sclerosis. My father sometimes speculated that the Japanese Imperial Navy had killed the man after all, that some of the shrapnel my grandfather carried inside him from that long-ago fight in the Pacific was toxic, and slowly poisoned him over the years.

Bruce I retired from the Navy in September 1952 at the age of forty-one, due to his physical disability. In recognition of his service record, and according to naval practice, he was advanced one grade from captain to rear admiral, which made him, in the Navy's grim parlance, a "tombstone admiral." As the doctors had predicted, but could do nothing to prevent, his paralysis advanced and in the end left him bedridden and despondent—a cruel end for a gallant man. My dad never spoke of his father's later years, which were difficult and unhappy. But the story of Bruce I's experience at Guadalcanal—persevering through chaos, concussion, and long odds to an unlikely triumph—was, I think, the most important narrative in my dad's life. Getting knocked down was inevitable. The real test was whether you got back up. Tight-lipped about his own experiences, my father was generally voluble about those of his parents. Whatever else was going on in our lives, he enjoyed talking about my grandfather Bruce and the dark early days of World War II.

The McCandless family moved to California in the summer of 1952. Two years later my father graduated fifth in a class of 855 from Woodrow Wilson High School in Long Beach. Long Beach Wilson, as it is called, is a formidable institution that stands just a mile and a half from the Pacific Ocean, the kind of expansive California secondary school that even in the fifties boasted golf, water polo, and swimming and diving teams. One of Bruce's teachers at the time recollected that the young man was "always confident and ambitious," and that he spoke even then about wanting to walk on the moon. In his senior year his grades were perfect, save for a single B in PE. His English teacher, Mrs. Helen Minami, gave him an A and wrote to Admiral and Mrs. McCandless, "You are fortunate parents."

I know precisely one story related to those days. It concerns a road trip my dad and a couple of buddies made to Indio, where, according to him, they drank date milkshakes. This is not the stuff of adolescent legend. I suspect he simply didn't have many high school stories. Only sixteen when he graduated, Bruce was a slender, studious kid set down like an alien life-form among the bronzed blonde surfers and hot-rod enthusiasts of Southern California. It couldn't have been an easy transition for him. And yet, broadly speaking, it was a great time to be a teenager.

America in the fifties was the undisputed leader of the free world. The country was flourishing, optimistic, triumphal even. In 1952, the same year the stock market finally climbed back up to pre-Depression levels, unemployment fell to 2.7 percent. American manufacturers of airplanes and turbines, televisions and refrigerators were heavyweight champions of the global free market. For car maker General Motors, for example, "it was a brilliant moment, unparalleled in American corporate history. Success begat success; each year the profit expectations went higher and higher." America was growing cleaner, richer, and more scientific by the day. Ray Kroc was measuring the circumference of the sliced pickles his McDonald's franchisees put on their hamburgers. Holiday Inn built its first dozen identical, spotless roadside hotels in the South, and Swanson introduced the "TV dinner"—a geometry project you could eat. The world wanted Coke and Cadillacs, Western movies and Westinghouse motors. The United States was happy to oblige.

Bruce McCandless breathed in the buoyancy. It became part of him, manifesting as a deep-seated belief in himself and his brain and the clean-shaven, odorless righteousness of American life. At Wilson High he was Phi Beta Kappa and a member of the German Club. He won the Southern California Chemistry Prize in 1953 and served on Wilson's all-school dance committee, a position that presumably involved little to no actual dancing, which he didn't like and couldn't do. He was also a student ambassador and a wannabe diplomat. One photo from these days shows him barefoot and wearing some sort of military cap, preparing to enter a mock United Nations assembly as a delegate from the People's Republic of China. He may have toyed with the idea of a career in the Foreign Service, but by graduation day his course was set. Just as his father and grandfathers before him, he'd won an appointment to the Naval Academy. One of his classmates wrote in his yearbook, the *Carillon*: "To the brain of the senior class. Congratulations on your appointment." And another: "Wishing you luck at Annapolis. Do your best for America's rockets and missiles!" This was clearly his intention. On May 15, 1954, young Bruce acquired George P. Sutton's *Rocket Propulsion Elements: An Introduction to the Engineering of Rockets*, which included chapters on topics such as "Nozzle Theory and Thermodynamic Relations," "Rocket Propellant Performance Calculations," and, as would be used in the shuttle program a quarter-century later, "Solid Propellant Rockets."

IN THE SUMMER OF 1954 Bruce McCandless boarded the Atchison, Topeka & Santa Fe's Super Chief for the trip from Los Angeles to Chicago, where he switched trains to ride the Baltimore & Ohio's Capitol Limited to Washington, D.C. It was an unusually hot summer, especially in the Midwest, but to the slender, cardigan-wearing teenager the western world seemed to be firing on all cylinders. Dwight D. Eisenhower was a sensible, unflappable president, as sturdy as a pair of boots. A friend of the boy's grandfather Bradley, a congressman named Richard M. Nixon, had been plucked from political obscurity to be Ike's vice president and seemed to be doing just fine. In May the Supreme Court issued its landmark ruling in *Brown v. Board of Education*, signaling the nation's intent to address longstanding racial disparities in its public education system. The Korean War was over, and America had successfully tested the hydrogen bomb, the weapon the Pentagon thought would guarantee the nation's military supremacy for the foreseeable future. Doris Day, Kitty Kallen, and the Four Aces, all good-naturedly wistful and as clean as showroom Chevrolets, were radio stars that spring. Little did they know that in just a few weeks, a teenage truck driver named Elvis Presley would step into a Memphis recording studio and take a pickax to popular music.

As his train rumbled eastward, young Bruce, sixteen years old, dark-haired and blue-eyed, five foot nine inches and all of 142 pounds, grew increasingly excited. He knew he would shortly be required to take an oath of service. He would swear to defend the Constitution against all enemies foreign and domestic, and he felt a mixture of appreciation and apprehension as he followed the cross-country route on his pocket atlas, checking off the names of the cities and towns—Barstow, Santa Fe, Dodge City, Topeka—he went through. In early June, in Annapolis, where his father and two grandfathers before him had studied and marched and done 10,000 push-ups, he joined a gifted class of "plebes"—short for plebeians, as entering midshipmen were sneeringly called.

The Academy was then and remains today a demanding place. Founded in 1845, it has long supplied many of the nation's most accomplished naval officers, including Chester Nimitz, William "Bull" Halsey, Raymond Spruance, and Hyman Rickover. It has also produced more astronauts than any other institution of higher learning. Among Bruce McCandless's classmates on those green fields were Chuck Larson, a future admiral and commander in chief of the Pacific Fleet; a "sturdy conversationalist and party man" who would spend several years being tortured by Vietnamese communists and survive to become Senator John McCain; and a studious, fiercely intelligent young midshipman named John Poindexter, who later served as an important policymaker and advisor to President Ronald Reagan.

There were no such distinctions apparent that summer, though. All plebes underwent the same grueling regimen of physical testing and cultural indoctrination. Their locks were shorn. They were issued a little book called *Reef Points* and required to memorize the uniform chain of command, the profiles of various naval vessels and airplanes, and such nuggets of wisdom as

Teddy Roosevelt's "The Man in the Arena." When asked at the dining table how much milk was left in the pitcher, they learned to respond with: "Sir, she walks, she talks, she's full of chalk. The lacteal fluid extracted from the female of the bovine species is highly proficient to the nth degree."

In September 1954 young Bruce wrote to his grandmother Sue Bradley about the indignities of being a plebe. The upperclassmen had just returned to Annapolis, he said, and were "making life rough on us . . . We have to eat all our meals sitting on only three inches of our chairs. We must also eat at a 'brace,' a sort of sitting position of attention. Then the upper classes take pride in asking us 'professional' questions. (What does the USS *Iowa* weigh? 45,000 tons? No!)" The plebes marched in close-order drill. They beat each other senseless in the boxing ring. They knotted and spliced, learned marksmanship and how to salute, were berated by upperclassmen, sent on pointless errands, and taught the rude mechanics of military discipline. In short, they started becoming naval officers. And then the school year began, and things got serious. The academic requirements were rigorous. Failure was a lurking fear. And all the while the plebes faced constant assessment of their deportment, discipline, cleanliness, and conformity—their *grease*, as it was called. This was hard on the brigade's free spirits. Aside from his academic struggles, which he himself acknowledged, Johnny McCain faced bilging out—expulsion, in other words—for demerits on more than one occasion.

Bruce McCandless found his true self at the Naval Academy. For a kid who loved absorbing facts and formulas, who thrived on structured learning and intellectual competition, it was a kind of secular paradise—Sparta, with slide rules. Assigned to 19th Company, Bruce was described by the Academy yearbook, the *Lucky Bag*, in his senior year as "Mac, electronics wizard extraordinary. Using a scientific approach in all fields, [he] stood first in the class academically Plebe year and has remained at the top ever since." Though he received just enough demerits to knock him down to number two in the class of 899, behind John Poindexter, he was a brilliant and enthusiastic student, rumored by his classmates to have an unfair academic advantage because he had two brains.

Bruce wasn't shy about demonstrating his talents. Decades later his classmates recalled that he was notorious for taking his calculus and physics exams in pen, rather than pencil, because he never felt the need to correct himself. Annoyingly, he still finished before everyone else. He made lifelong friends of his classmates Ernie Merritt and Dave Willingham. He joined the photography club. And while his marching cadences weren't as precise as they could have been, and he was never a fan of physical exertion for its own sake, he nevertheless found a way to excel outside the classroom, winning a varsity letter for his participation on the Naval Academy's sailing team.

At the Academy—Canoe U, the middies called it—he thought about becoming a submariner. He was impressed by the visit to Annapolis of the world's first nuclear submarine, USS *Nautilus*. Sleek, black, and powerful, longer than a football field, the *Nautilus* was a wonder. Named after the vessel in Jules Verne's novel *20,000 Leagues Under the Sea*, the ship's emblem had been designed by The Walt Disney Company, further evidence of its fantastic pedigree. In August 1958 the *Nautilus* accomplished the astonishing feat of traveling *under* the North Pole.

For an aspiring engineer like Bruce McCandless, submarining must have seemed like a tantalizing opportunity—high tech in the service of high adventure. But like many other young men of his generation, he was shaken by the launch and successful flight of a Soviet satellite. Sputnik-1 was an unassuming metal orb, roughly the size of a beach ball, that trailed a set of whiplike antennae like a protozoan with four flagella. The Soviets sent it into Earth orbit on October 4, 1957, whence it emitted recurrent beeps via radio. My dad, an officer of the Academy's radio club, hastily rigged a wire antenna to the Hallicrafters SX-71 shortwave receiver in his fourth-floor room. He dropped the wire out the window all the way down to the "moat," the sunken cement walkway around Bancroft Hall, the enormous dormitory where the midshipmen lived. Sure enough, he and his buddies could hear the taunting tones of the Soviet satellite as it passed overhead. ("Beep. Beep. Beep," Dad would later remember. "That's Russian for *We will bury you*.") They listened off and on for the next three weeks, until the beach ball's silver-zinc batteries died.

Sputnik eventually fell back into Earth's atmosphere, burning up on its way toward the planet's surface. There wasn't a whole lot of science to it by modern standards, but the effect of Sputnik's flight on the American public was electrifying. RUSSIANS LAUNCH FIRST SPACE SATELLITE; CIRCLING EARTH AT 18,000 MILES AN HOUR, announced *The Washington Post*. THE SPACE AGE IS HERE, trumpeted New York City's *Daily Express*. The Russians had managed to supplant the Germans and the Japanese as America's sociopolitical bogeymen soon after World War II. Watching their aggressive brand of authoritarian communism, and a military that during the course of the fifties had managed to develop atomic and hydrogen weaponry to match the destructive power of the West, the United States had come to believe it was locked in a cage match for control of the globe with the soulless fanatics of Moscow. Most Americans believed their country was bound to prevail. Sputnik, then, was a slap in the face—a highly visible and embarrassing rip in the nation's casual assumption of scientific preeminence. There was now a socialist sentry on the celestial shore. With one swift stroke, the Red Menace had established a significant and very obvious lead in the technologies of surveillance, orbital mechanics, and space warfare.

Nor was the satellite's flight the only blow. The Gaither Report, compiled by a blue-ribbon panel of experts appointed by President Eisenhower himself and issued just after Sputnik's flight, warned of the USSR's expansionist tendencies and ever-increasing nuclear missile—that is, rocketry—capabilities. On October 6 the *Post* ran the following banner headline: SATELLITE FLASHES PAST D.C. 5 TIMES; TRACKERS FAIL TO FIND IT ON 6TH TRIP; RUSSIANS MAY HAVE 'ULTIMATE WEAPON.' And then, on October 10: PRESIDENT PROMISES SATELLITE; MISSILE PROJECT SPEED-UP DUE; RED 'MOON' STILL GOING STRONG.

America's space agency, the National Aeronautics and Space Agency (NASA), was established the next year, in 1958. It inherited the work, employees, and budget of the National Advisory Committee for Aeronautics (NACA), assumed control of certain space-related projects formerly under the auspices of the Defense Department, and took over direction of the Jet Propulsion Laboratory at the California Institute of Technology in Los Angeles. Washington had for years been content to leave rocketry research to the armed forces, particularly the Air Force. No more. Politicians of all stripes agreed that it was time to make the space race a national priority, under centralized supervision.

Wernher von Braun and other former Nazi scientists and rocket engineers, "prisoners of peace," as they called themselves, had already been released from their holdfast in the Chihuahuan Desert. Now von Braun was given a seat at the nation's conference table—a wise decision, as it turned out, as von Braun's rockets would eventually supply the propulsive muscle, the great greasy explosive chariots of fire, that would carry Americans to the moon. NASA began drawing up plans to get the United States into a contest it hadn't previously even realized was underway. The aim was to equal and surpass the Soviets in every facet of space travel. Soon NASA would start looking for the men who could make it happen. They would be called *astronauts*.

My father was determined to be one of them.

4

YOU ARE ON FIRE

Bruce McCandless graduated from the Naval Academy on June 4, 1958, his future as bright as the glint of sunlight on his ceremonial saber. He won special prizes in electronics, electrical engineering, and practical navigation, and his academic achievements earned him the privilege of receiving his diploma from the hand of President Eisenhower at the commencement ceremony in Annapolis. Ike, balding and avuncular, commended the assembled midshipmen that day for their sense of duty. He encouraged them to serve the nation with honor. He also reminded them to serve with a sense of humor—which, he pointed out, communists were not allowed to have. After the men officially graduated and were sworn in, *en masse*, as junior officers in either the Navy, Marine Corps, or Air Force, the ceremonial cap tossing began. Nine hundred white flying saucers soared above the throng of elated young ensigns and lieutenants.

Sputnik's flight had rekindled my dad's dreams of walking on the moon. Though the career path to becoming an astronaut wasn't clear, since there weren't any astronauts yet, becoming a pilot seemed like a more logical starting point than submarining or serving aboard a surface ship. Accordingly, Ensign McCandless reported in the fall of 1958 to the Naval Air Station in Pensacola, Florida, for basic flight training. In January 1959 he completed the Navy's primary training syllabus at Saufley Field, having made his first solo flight in a T-34 Mentor aircraft. He graduated from the Naval School of Pre-Flight that year, standing first in a class of twenty-two, even as he took a Russian language course at night at Pensacola Community College. He finished his course of instruction at the Naval Air Station in Kingsville, Texas, in 1959. After making six solo landings aboard the carrier USS *Antietam* while it was underway in the Gulf of Mexico, he earned his wings in March 1960, when he was officially designated a United States Naval Aviator—the Navy's term for a pilot. He then relocated to Memphis, Tennessee, for a course in naval aviation maintenance from the Navy's Technical Training Command before further instruction in carrier-based aircraft.

In Memphis he met my mother, Alfreda Bernice Doyle. Having pronounced my mother's magisterial first name, so redolent of ancient Anglo-Saxon glories, let us now do as she would have asked and forget that it existed. She was always embarrassed by her name—the Alfreda more than the Bernice—and preferred to go by Bernie instead. Born into a middle-class family in Roselle, New Jersey, on April 24, 1937, Bernie grew up in a mixed neighborhood where people still congregated by ethnicity—the Italians and the Jews, the Irish and the African Americans. It was mildly adventurous of my mother to crush on an Italian boy. While her family name was Irish, the family heritage was mostly Dutch. ("The Irish and the Dutch," she used to repeat, referencing a taunt she heard as a kid. "They don't amount to much!") I like to think of her hometown as Springsteen territory, though Springsteen, also a product of Dutch and Irish ancestry, is twelve years younger. Roselle is a suburb of Newark, but the social and cultural magnet then as now was nearby New York City, center of the universe and the epitome of wealth, style, and glamour.

My grandfather Charles Doyle was a tall, bald, mild-mannered individual who played minor league baseball. When his athletic dreams didn't work out, he went to work at a local Singer sewing-machine plant and stayed for the next fifty years. I have the gold watch he was awarded by the company to prove it. It was while working at Singer that he met my grandmother Jeanette. She was twenty-three years his junior, a tease and a lover of lyric poems, full of dreams and elaborate but generally impractical plans.

According to my mother, Jeanette set her sights on Charles and pursued him with a frankness that was unusual, indeed slightly scandalous, for the day. But her plan worked. The couple married in 1928 and had five children. The firstborn of the family, a girl named Madeline, died shortly after birth. Mom had two older sisters, Janet and Virginia, and an older brother, Chuck, or, as he came to be called, Brud (for Brother). Bernie was a shy, melancholy child with blonde hair and blue eyes. She loved her parents and often reminisced about them. I never knew my grandfather Doyle, but my mother described him as a gentle man who was always kind to her. He was sixty years old when she was born. Perhaps he was bemused by the experience. Certainly he would have been fatigued. From him she absorbed a devotion to the Brooklyn Dodgers and the Friday night fights on Gillette's Cavalcade of Sports, and though she could rhapsodize about that legendary Marciano-Walcott title bout in September '52, she could never actually watch more than five minutes of a fight before getting upset about the violence. Despite showing an occasional wild streak, her mother was a devout Methodist and generous to a fault. The Doyle house was known as an easy mark for passing hobos, who could count on receiving some snack—a sandwich, say, or a piece of fruit—from my grandmother Jeanette even in the darkest days of the Depression.

While the Doyles weren't as intimately involved in World War II as the McCandlesses, they were equally patriotic. One of my mom's favorite recollections involved swapping stories about the war with her siblings and cousins. Her father was an air-raid warden, charged with patrolling the streets of Roselle to enforce the lights-out edicts meant to confound Nazi attempts to bombard crucial targets in suburban New Jersey. Though she wasn't always clear on the details, when the war news was good, Bernie would stand up, raise a fist, and shout, "That's what you GET, Hitler!"

The family went to the Jersey Shore every summer for two weeks, and in my mother's memory these times were golden and untouchable. The Doyles crowded the boardwalk at Ocean Gate, her father resplendent in a straw hat, white shirt, and suspenders, her mother wearing a cotton print skirt that billowed in the breeze. The kids played arcade games and listened to the caws of the gulls. They ate steamed crabs, sampled root beer and saltwater taffy, and lounged in the July sun. Like other Jersey natives, Mom was tribalistic and intensely loyal. She never quite forgave Elizabeth Taylor, for example, for breaking up Eddie Fisher's marriage to Debbie Reynolds. And like thousands of American teenagers, Bernie grew up following Frank Sinatra and Marlon Brando, Ingrid Bergman and Audrey Hepburn. Her true love, though, was Elvis.

Mom graduated from Roselle's Abraham Clark High School in 1955 and attended Ohio's Baldwin-Wallace University, where she was a Phi Mu, for two years. She left college without taking a degree, rudderless and vaguely dissatisfied. She was an indifferent student, she admitted later, more interested in marriage than a career. In 1959 she moved down south to live with her sister Janet, who was attending college at what was then called Memphis State University. Bernice Doyle was beset throughout her life by spells of lassitude. I think of her during this period as someone who spent a lot of time at the movies, whether she was in a theater or not. She was a lifelong romantic, an incorrigible daydreamer with an easy smile that masked her many insecurities.

One afternoon in May 1960, Bernie rented a horse from a local stable and went for a ride. The horse was spooked by a passing car and refused to cross a road to get back to the stable. Fortunately, Mom was assisted by a Navy pilot named Joe Hart, who happened to be passing by. Hart went and rented a horse for himself, and Joe and Bernie rode together awhile. Following the ride, Bernie met Joe's roommate, another young aviator, named Bruce McCandless. My dad eventually asked Joe if he could ask my mom out on a date—"as if I was somebody's property," Mom later sniffed. Joe said sure, and things happened quickly after that.

My parents' first outing was a flight on a rented Cessna 172. "The boy's bedazzled!" my grandfather remarked when he saw Bruce and Bernie together. The young couple were married on August 6 in the Naval Academy Chapel in Annapolis. My cousin Jo Thomas was there. A teenager at the time, she grew up to be a Pulitzer Prize–winning reporter for

The New York Times. She remembers the wedding clearly. It was a glamorous event, she recalls. And Bernice McCandless was the most beautiful woman she'd ever seen.

SHORTLY AFTER THE WEDDING, Bruce McCandless reported to Naval Air Station Key West for weapons systems and carrier landing training in the Douglas F4D-1 Skyray. The Skyray was the Navy's front-line fighter, a nimble aircraft specifically designed to intercept and engage enemy planes at high altitude. Its name came from the designers' notion that the prominent delta-shaped wings suggested the shape of the manta ray, but in flight the plane actually looked more like a swallow, its compact fuselage dwarfed by those big wings.

These months in Florida were a precious time in my mom and dad's marriage, the time all others were measured against. Pre-cruise-line Key West was a sleepy place. When they could, the young couple swam, sunbathed, and explored the little curio shops along Duval Street. Their letters back and forth for the next several years return to the little island like the chorus of a pop song, and for the rest of their lives, Bruce and Bernice kept a polished conch shell—King Konch, they called it—as a souvenir of their sojourn in the almost-Caribbean.

Unfortunately, the honeymoon was short-lived. "Bruce certainly did well in the training," wrote his commanding officer in a letter to my grandfather McCandless. "His carrier qualifications in the F4D were the best we've ever had." After two months of training with Attack Squadron 43, Dad was assigned as a replacement pilot to Fighter Squadron 101 (VF 101), known as the Diamondbacks. He went to sea on the Sixth Fleet's aircraft carrier USS *Forrestal*, the Navy's so-called supercarrier, a ship that boasted a flight deck of some four acres. In December 1960 he was transferred to VF-102, the Grim Reapers. Between 1960 and 1964 Bruce McCandless flew the Skyray and, later, the new McDonnell-Douglas F-4B Phantom jet, a sort of flying fist. Bigger, faster, and more lethal than the Skyray, the Phantom was fifty-eight feet long and had a wingspan of thirty-eight feet. It weighed more than 55,000 pounds when armed and fully fueled but could travel at twice the speed of sound, proving, some observers said, that even a brick could fly if it was attached to a large enough engine. Heavier and less maneuverable than its Soviet counterparts but faster and more powerful, the Phantom was the West's premier fighter plane for decades afterward. It was also remarkably durable: Some of the planes remain in limited service in various countries even today.

It didn't take long for Bruce McCandless to test his training. On a bright, cool morning in February 1961, with visibility at five miles and an air temperature of 56 degrees, he tried to land his Skyray on the *Forrestal*. The ship's arresting wire not only failed to restrain the Skyray, it actually tore off the tail end of the plane. Dad was twenty-four years old, a newly minted naval aviator on the second day of his first Mediterranean cruise. The Skyray was slowed by the wire but not stopped, and now the far end of the landing deck was coming up fast. *Too fast.* The young lieutenant didn't know exactly what had happened, but he realized the generally reliable system for bringing his aircraft to a stop had failed. He also knew what came next. Unless he was extremely quick and a little bit lucky, he was going to die.

He slammed the Skyray's J57 engine into afterburner and waited several agonizing moments—the sea coming closer—until it kicked in. Then he willed every ounce of the plane's 16,000 pounds of thrust as he bolted, with nothing between his metal cocoon and the sea but a rapidly diminishing slab of steel deck. The aircraft barely managed to make it off the carrier. Fortunately, the Skyray was very good at the one thing that was necessary now, *climbing*, and Dad manhandled the plane up to 500 feet. But he couldn't relax. A voice from the *Forrestal* crackled in his ears. *Pull*

up and get out. Eject! You're on fire. Repeat. YOU ARE ON FIRE! Bruce
McCandless glanced back over his shoulder. Sure enough, a fifty-foot
plume of flame was trailing from the Skyray. Pointing the plane out over
open sea, Dad tried to deploy the ejection seat's face curtain—the mechanism that dropped down over a pilot's head to protect him when leaving
the aircraft—so he could eject, but the mechanism didn't work. The curtain was situated above his head and slightly behind him, with a handle
attached for activation. He pulled the handle three times without success.

"At this point," the novice pilot later deadpanned, "I wasn't exactly sure
of my next course of action." He noticed that his crippled Skyray had
banked to the left and was nose down at about 500 feet. He leveled the
plane's wings. He forced the nose up, trying again for altitude, and came
out of afterburner, thinking this last action might help to put out the fire
or at least keep the aircraft from exploding and killing him for a few seconds longer. Without the option to eject, he was going to have to roll the
Skyray and bail out over the side, hoping as he did so that he wasn't cut in
half by the plane's tail fin as it went past. It was an ugly option, but not one
that he had much time to think about. He jettisoned the plane's canopy
and noticed the cloth portion of the face curtain handle whipping about
in the wind. He grabbed it, worked his hand up the handle itself, and
yanked again. Yanked *hard.* Ejection seats—safety devices, of a sort—are
not actually very safe. They are among the most complicated mechanisms
on a fighter jet, and involve use of a controlled explosion to propel the pilot
out of and free from the aircraft. Stories of injuries and deaths incurred as
a result of ejection abound. Chuck Yeager reportedly called ejection from a
jet airplane "committing suicide to prevent getting killed." Even mechanics
servicing planes on the ground have been killed by accidentally triggered
ejection seats, such is the violence of their propulsive mechanisms. But this
time it worked. The curtain came down, and the ejection seat fired.

The young lieutenant felt nothing, but he realized he was tumbling violently in midair. Good news, this. He was out of the plane. His parachute
deployed automatically, and the tumbling stopped. But as he drifted down,
Dad glanced up and realized he was now in the flight path of another

Skyray approaching the carrier. He'd escaped from a death by immolation or explosion only to face being smashed like a bug on a windshield. He frantically waved the pilot off, and the aircraft banked away. In the meantime, Dad's plane plunged, still burning, into the sea.

The ocean came up as flat and hard as a cement floor. Bruce McCandless was three feet under the Mediterranean by the time he realized he'd made impact. When he fought his way back up, he was relieved to see that his chute had collapsed and now floated like a dead sea creature on the surface not far from him. This was important, because it was not unknown for a parachute to billow in the wind and drag a pilot through the water until he drowned. He disconnected himself from the parachute lines, the last of which he had to cut with a knife. Free of the chute and buoyed by his life vest, McCandless surveyed his surroundings. He was astonished by how far he'd traveled from the carrier. The ocean, which had seemed so flat to him while descending, was actually quite choppy, and he spat out another mouthful of salt water as he rose and fell with the swells. But it was okay. He was breathing. In the distance he could see the plane guard helicopter, a dual-rotor HUP-2 Retriever, heading his way. He rolled his neck from side to side, kicked his legs, assured himself he was healthy. Already some part of his engineer's mind was wondering what had happened to send him on this terrifying package tour of his own mortality, but he told himself not to think about it now. Bruce McCandless closed his eyes for a moment, grateful for the bracing chill of the sea, the pastel tints of the morning sky. Not long afterward he hauled himself into the rescue seat dropped by the helicopter and was hoisted by winch up through a hatch in the chopper's floor. The young lieutenant had survived unscathed, with only a rip to his MK4 exposure suit to show for the experience. It was as close a brush with death as an aviator could have.

Unfortunately, the ordeal wasn't over. The Navy commenced an investigation of the incident, which included taking the statements of both Lt. McCandless and his commanding officers. The central question, obviously, was what had caused the accident. Was it a material or mechanical failure or was it rather some mistake made by the pilot, such as trying to land at too high a speed? A finding of pilot error might have been disastrous to a young lieutenant's career prospects. In June 1961, four months later, the inquiry concluded that the Skyray's tailhook and attached parts had separated from the rest of the aircraft on landing as a result of a previously undetected crack in the body of the plane. The Navy therefore found material failure as the "single contributing factor" of the crash.

Bruce McCandless was exonerated.

WHEN I WAS GROWING up, my father was reluctant to discuss this event, although of course it was exactly the sort of pulse-pounding near-disaster I wanted to hear about most. It couldn't have been a pleasant memory for him. The golden-boy aviator, blessed with a brilliant heritage and a budding career, lean, ambitious, and marked for success, suddenly found himself balanced on the razor-thin edge between life and death. He'd worked hard all his young life. He had a beautiful wife waiting at home, plans to start a family, a hundred ideas in his head. So why him? Why had the metal on the tail section of *his* aircraft snapped on this particular day?

The answer, of course, is that there was no good reason. It was sheer dumb luck, just like the sheer dumb luck that had killed a million other lean, ambitious, gifted bastards since the world began. And so as Bruce McCandless tumbled through the sky on that bright February morning, seeing nothing but ocean and sun, ocean and then sun and then ocean again as if he were strapped to some turbocharged medieval wheel of fortune, he had to have been wondering what it all meant. Was it worth chasing a dream, however noble, if it meant that *you*, an honor student from Woodrow Wilson High School, a talented ham radio operator, might end up at the bottom of the Mediterranean, imprisoned pale and forever unspeaking in your strange flying machine amid the wreckage of Phoenician triremes and Moorish gallivants, companion to the eyeless, bronze-helmeted dead? Virtue was no shield. Purity of heart hadn't saved young Quentin Roosevelt, scion of the famous political family, when he was shot down and killed while fighting for the cause of liberty in the skies above France in 1918. Sheer physical vitality meant nothing on a day in July 1861 when David McCanles walked into the office of the Overland Stage Company to collect a debt and was shot dead by the Redheaded Gunman. Dave was a big man, as healthy as a high wind, but he had no immunity to bullets. He staggered backward into the yard, blood welling up from his wounds. He locked eyes with his young son for just a moment before he lay back in the dust. A twist of fate. A sudden calamity. Brilliance and industry and intelligence couldn't preserve the lives of other men undone by simple mistakes, perilous odds, aces and eights. They fell into that great black hole, the vast unexplainable nothingness, and we never learned their names, or what they had to offer the world.

Do astronauts, or proto-astronauts, think this way? Do they wish they could re-roll the dice? Choose some simpler, less hazardous life? Or do they just run through the checklist, as my dad said he did, and brace for impact? I don't know why he was reluctant to talk about the experience. Death-defying escapes are the currency of aviator legend, and narrative is among the easiest ways to blunt and redirect personal terror. Maybe he just wanted to bury what had to have been the most harrowing three minutes of his life and move on from the achingly slow investigation that followed.

Then, too, the incident put my dad in an unaccustomed situation. The *Forrestal*'s newspaper of record, the *Antenna*, printed this account of the crash on February 21, 1961: "Lt. Tom Anderson and his crew of Helicopter Utility Squadron Two performed an outstanding rescue when they returned Lt. (jg) Bruce McCandless of VF-102 safely aboard *Forrestal* within minutes after his plane crashed into the sea yesterday." The idea that he had been rescued wouldn't have sat well with my father, who prided himself on being in control of all situations at all times. It might have been especially galling to read a story like this on a day when the *Antenna* also reported on stress tests being performed on a prototype of NASA's new Mercury space capsule. Bobbing in the ocean in the prop wash of a rescue helicopter's rotors was not where Bruce McCandless wanted to be as the nation's space program inched forward.

DAD WAS AT SEA when I was born in Annapolis on August 15, 1961. Tracy arrived two years later, on July 13, 1963, at the naval hospital in Virginia Beach, Virginia. Virginia Beach was a tough billet for Bernice McCandless. My father's tours of duty with the Atlantic Fleet were so lengthy that my mother was essentially a single parent. She had a tiny house in Lynnhaven, two infants to take care of, and a rambunctious German shepherd to wrangle. But she was young and in love, and she made the best of it. One of my dad's shipmates on the *Enterprise* was Johnny McCain. Because their names were so similar, my father occasionally wound up with some of McCain's laundry, which was washed and packaged aboard ship by the crew. When the *Enterprise* was in port, the young aviators caroused together. My mom delighted in sidling up to the handsome McCain at parties and whispering, loud enough for those nearby to hear, "Johnny, I still have a pair of your underwear at my house."

Bruce McCandless was aboard the *Enterprise* during the ship's participation in the American blockade—called a quarantine, for diplomatic reasons—of Cuba during the Missile Crisis in 1962. The confrontation occurred when the Soviet Union attempted to install nuclear-armed medium- and intermediate-range ballistic missiles in the island nation, well within striking distance of American cities on the country's East Coast. The precipitating factor appears to have been the U.S. decision, earlier that year, to station American missiles in Turkey, relatively close to the Soviet Union. Whatever the justifications, the placement of missiles in Cuba raised alarms in Washington.

President Kennedy deployed military assets, including the *Enterprise*, the world's first nuclear-powered aircraft carrier, to stop any further Soviet efforts. There was an obvious mismatch of forces in the Caribbean. The "Big E," almost as long as the Empire State Building is tall, with more than sixty warplanes aboard and a crew of 4,500, was capable of wiping Havana off the face of the earth all by herself. But Castro and his *Fidelistas* were seasoned fighters, dedicated to their revolution, and they'd made short work of Kennedy's invasion force in the Bay of Pigs disaster a year and a half earlier, killing or capturing something like 1,600 American-backed insurgents. They weren't likely to throw up their hands and surrender at the sight of the Stars and Stripes. The standoff had global implications, and for a time it looked as if war was imminent. American kids did duck-and-cover drills at school. Supermarkets were ransacked, shelves emptied. My wife's parents, Joe and June Fuller, decided to move the 150 miles from Houston to Nacogdoches in East Texas, to get away from what they perceived as a target-rich environment for missile attack.

As the confrontation wore on, my father and his squadron mates flew patrols in enemy air space, ready to fight off any threats from the Cuban military, which was known to have surface-to-air missiles and several dozen Soviet MiG-15s and MiG-19s at its disposal. On October 22 Dad wrote to my mother from the *Enterprise*. The script is the same as that in his other letters home, and the lines on the paper are straight, but his writing is larger and his pen strokes bore into the paper. His heart rate has clearly been elevated by the threat—or promise—of action:

> *From all indications, tomorrow is the day most likely to see the outbreak of hostilities. If Cuba cooperates, the missiles go back to Russia and we come back home. If not—I gather that we have to take the island foot by foot . . . I don't particularly like staring war in the face, but I'm glad to see that we as a country are no longer idly sitting by while Castro does whatever he pleases. I'm now carrying two pistols with about 100 rounds of assorted ammunition and a load of other paraphernalia too numerous to mention.*

While my father no doubt intended to be reassuring, the image of him festooned with bandoliers, two pistols, and a fighting knife only heightened my mother's fears. Fortunately, though, war was averted. The Soviets withdrew their weaponry, and, confidentially, Kennedy agreed to withdraw American missiles from their objectionable locations in Turkey and Italy. In the end, there was only one combat casualty of the crisis, an American U-2 pilot named Rudolf Anderson, who was shot down by a Soviet-made missile. On October 27 President Kennedy promised that the United States would make no further attempts to invade Cuba. The crisis ebbed, and the world breathed a sigh of relief. By November 8 the Soviet missiles were on their way home, and life in the States was returning to normal. Richard Nixon had just lost his race for governor of California and promised he was done with politics. "You won't have Nixon to kick around any longer," he told the press—erroneously, as it turned out, as the press would find occasion for plenty more kicking. Former First Lady Eleanor Roosevelt passed away. Notorious Texas con man Billy Sol Estes was convicted of fraud in a Tyler court and sentenced to eight years in prison. Despite a return to the customary rhythms of American life, it was hard to avoid the fact that the Cuban Missile Crisis was as close as the United States and the Soviet Union ever came to converting God's green earth into something that looked a lot more like an asteroid.

ONE OF THE OLD sayings about war is that it consists of long periods of boredom punctuated by moments of sheer terror. Peacetime seems to have felt much the same way to Bruce McCandless. When he wasn't exploding out of his airplane or strapping pistols to his survival suit, the young aviator pecked out a whole library of letters on his little Underwood typewriter. Fellow officers banged on the bulkhead to complain about the racket, but he clattered on. He spent most of his time overseas in the Mediterranean. His missives to my mother are typically captioned by date and the statement AT SEA, with few details about what he is doing other than flying. He asks for news and photos. He discusses the couple's jury-rigged budget and details day trips in Corfu and Cannes and Genoa. He obsesses over an ignition problem with the Triumph TR-3 he left at home and fantasizes about establishing a private radio circuit that he and my mother could use to talk whenever they chose. Two letters mention that he is occasionally sealed up in a pressure suit and helmet and displayed as an astronaut (or "man from outer space," as he puts it) for tourists visiting the ship on port calls. On September 2, 1962, in Golfe Juan on France's Cote d'Azur, Bing Crosby and his entourage came aboard the *Forrestal*. Movie star Crosby, the man who recorded the most popular single in history, "White Christmas," was as dapper as always. He posed for a photograph with the pseudo-spaceman, who noted, unimpressed, that Der Bingle was getting old. In other letters the young lieutenant bemoans the meagerness of his military pay and the Navy's arbitrary rules (no sitting on bunks while fully dressed) and regrets missing the births of his children. On one occasion he relates the following conflict:

> *I'm afraid I made an enemy or so this evening. I caught one of the pressure suit riggers with a stencil prepared for a batch of anti-Negro propaganda (an "application" for membership in the NAACP . . . definitely not what we need for good relations within the squadron) and turned him over to the XO. This cruise is not a "happy cruise!"*

Even as he dealt with the stresses of carrier flights and shipboard discipline, he searched for ways to challenge himself. On July 5, 1963, he wrote my mother that he had "borrowed a math book . . . and have been learning to manipulate determinants and matrices. Great sport! I hadn't realized it up until now, but their use greatly simplifies the solution of systems of simultaneous equations—and I have plenty of those around in the 'Transistors' correspondence course."

Engrossed in equations or not, life at sea as a junior officer was often difficult. Junior-grade lieutenants were at the beck and call of their commanders. On the *Forrestal*, Bruce McCandless served for a time as squadron duty officer and was evidently responsible for facilitating communications back to the States. One night, he wrote to my mother, a senior officer "came in slightly inebriated and decided to send a message [back to the U.S.] at

0115 this morning. Good grief! I didn't get turned in again until 0230. This is the sort of thing that is beginning to get irritating . . . However, it should improve greatly when I make [full] lieutenant this June. The Navy still has a lot to offer, but I wish I could be near you more of the time."

The problem wasn't the Navy. It wasn't even the separation from my mother, though that obviously hurt. The problem was boredom. The carrier landings, math books, and correspondence courses weren't enough. His mind needed more to do. He remained fascinated by electronics, and started looking for graduate schools where he could immerse himself in the field. He took the Graduate Record Exam and sent out applications, primarily to West Coast schools: Caltech, Stanford, Berkeley. He and my mother came up with a "war cry": *P.G. [post-graduate studies] in '63, or Out the Door in '64.*

Bruce McCandless flew his final carrier sortie early in 1964. He returned stateside and for three months served as an instrument flight instructor in Attack Squadron 43 (VA-43) at Naval Air Station Oceana, Virginia. In May of that year, he built a little wood-sided trailer. He and my mom packed up all their earthly possessions, tossed me and Tracy in the back of my dad's new Jeep, and drove the 3,039 miles to their new home in Mountain View, California. There, my dad was assigned to the Naval Reserve Officers Training Corps unit at Stanford University while he pursued graduate studies in electrical engineering.

I suspect my father was relieved to leave the flight line. As much as he loved his years at the Naval Academy, he soon wearied of the hierarchy and petty routines of active duty. He wasn't a daredevil. I never saw him drunk. He didn't smoke, or gamble, or even drive particularly fast. Flying fighter jets is a notoriously dangerous occupation. Add the dimension of operations at sea, and the danger increases; any aircraft carrier pilot has a dozen stories to tell about narrow escapes and tragic endings. One astronaut's older brother was a carrier pilot. On a routine takeoff attempt, he failed to achieve adequate launch speed and his aircraft plunged into the ocean just in front of the carrier, which then plowed over him, killing him instantly. And while failures could be catastrophic, even successful landings were traumatic, resulting in, among other things, a significant risk of spinal injuries. While my dad was proud of his time with the Diamondbacks and the Reapers, he was above all a rational human being. I think he was relieved to leave carrier service and move on to graduate school, where he could get reacquainted with his family and resume the real joys of his life: reading, thinking, and engineering.

At any rate, it was the right move. In July 1965 Air Force officers flying Phantom jets scored their first dogfight victory over a Soviet-made MiG-17 in the skies above a little-known land directly south of China. The land was called Vietnam. It would eventually become a cemetery for American aircraft, and the last resting place for many Air Force, Marine, and Navy pilots. War in Southeast Asia was coming, and with it winds of terror, sadness, anger, and change.

5

DON'T TREAT HIM
AS A PILOT

The McCandless family arrived on the West Coast in May 1964. My father began his postgraduate work immediately afterward, traveling to and from class on a purple Schwinn ten-speed, vanishing into his textbooks so completely that my mother sometimes said she felt like she was living alone. Stanford was a logical choice for Bruce McCandless's studies for several reasons. First, while the school had forged a formidable reputation in a number of fields, it was particularly strong in electronics research. For example, two brothers had developed the klystron (a specialized linear-beam vacuum tube) at Stanford in the 1930s, an important advance in radio technology. Hewlett-Packard, founded by two Stanford graduates, was located nearby, and Nobel laureate William Shockley, father of the semiconductor, the "man who brought silicon to Silicon Valley," had recently joined the faculty, though his behavior was becoming erratic, and he was increasingly interested in race-based theories of "dysgenics" that eventually cost him his professional reputation. In 1964 the university started its Stanford Linear Accelerator Center, a facility for study of highly energized atomic particles that was used by scientists who would go on to win three Nobel Prizes for their work.

The San Francisco area was a lure for other reasons as well. Not only did it have powerful symbolic resonance due to the first Bruce McCandless's service on the cruiser of the same name in World War II; it was also only a three-hour drive from the home of my dad's paternal grandparents, Byron and Mimi, in Mariposa to the east. Finally, NASA's Ames Research Center, home to the engineers who designed and controlled America's early interplanetary probes, was located nearby.

Not everyone at the university was aimed at the future. Some were trying to change the present. Oregon wild man Ken Kesey, for example, attended creative writing classes at Stanford and penned his novel *One Flew Over the Cuckoo's Nest* after taking part in experiments with hallucinogenic drugs at the Menlo Park VA Hospital under the supervision of Stanford researchers. In June 1964, not long after we arrived on the West Coast, Kesey set out from La Honda with Neal Cassady and a busload of like-minded others on the transcontinental acid trip that would herald the psychedelic sixties.

If my father noticed, he never said.

Bruce McCandless received his MS degree in electrical engineering on June 13, 1965, and paused for half an hour or so to enjoy the accomplishment. Then he plunged into his PhD. Now engrossed in the emerging field of plasma physics, he worked with, among others, the legendary Oscar Buneman. An Italian-born scientist of German ancestry, Buneman was briefly imprisoned as a young man by the Nazis for antifascist activities. Upon his release, he emigrated to Great Britain and helped to develop radar technology during World War II, thereby assisting mightily in Great Britain's ability to hold off the Luftwaffe during the darkest days of the conflict. At Stanford, he was known for his work in the fields of plasma electrodynamics, fundamental electrodynamic theory, and numerical analysis—and also for sleeping outside at night and for riding his bicycle to campus in "very brief shorts" on a daily basis. Dad completed the class work necessary for his doctorate by the spring of 1966. Professor Buneman described his new protégé as a "star pupil" and "an extremely competent man carrying an extremely heavy course load."

Unsurprisingly, my mother was lonely. She had no friends or family in California. Her job, she said, was to keep Tracy and me quiet as her husband churned through his engineering texts. My memories from the period are colorful fragments, not to be entirely trusted. In my mind, though, it is a sunny time, episodic and untroubled. I learned to love TV—especially reruns of *Adventures of Superman*, when the all-American hero from outer space was played by barrel-chested George Reeves, and *Mighty Mouse*, which, though intended as a parody, registered as purely heroic to me. My mother watched *The Secret Storm* in the afternoon as Tracy slept and I played with blocks. My dad took us to the drive-in to see *Thunderball*, in which James Bond makes a daring escape by means of a jetpack, and I remember sitting on the roof of the Jeep on a blanket, watching the inexplicable goings-on. Mom played records to cheer herself up in the mornings, and I did my best to help. As the Kingston Trio harmonized or some old album of military songs rang out, we'd sing and dance and go marching to Pretoria, or at least to Palo Alto. My father was dark and focused and generally on his way somewhere else. I was always surprised to see him. He had a long stride I couldn't keep up with and he hated noise. One day I gathered every model warship and jet airplane I owned and arranged all my plastic soldiers in battle lines to demonstrate that I was capable of projecting military power all the way to the dishwasher. To my eyes, it was an impressive show of force, and my way of announcing there was another power to be reckoned with in the house.

Dad looked on with amusement.

"Atom bomb!" he said, backhanding a pillow at my invasion force. He completely destroyed it. It was a sobering lesson, I suppose, in the advantages of technological superiority. But then he laughed, and that made it worth it.

Sᴘᴜᴛɴɪᴋ ᴡᴀs ᴛʜᴇ ʙᴇsᴛ thing that ever happened to the American space effort. The Soviets followed up quickly on their early successes, captivating the world by sending the first man into Earth orbit in 1961 and, two years later, the first woman, twenty years before an American female would duplicate the feat. These socialist exploits stung NASA, which was quite literally having trouble getting the U.S. space effort off the ground. Things finally started to work with the Mercury program, a series of six increasingly ambitious solo missions featuring members of the Original Seven astronauts. Alan Shepard rode a Redstone rocket into space and quickly back down again in 1961, and John Glenn orbited the Earth a few months later. On the final Mercury flight, in May 1963, Gordon Cooper circled Earth twenty-two times over the course of thirty-four hours. Cooper, a legendarily laid-back and occasionally unpredictable pilot, dozed off on the pad while awaiting launch and slept again while in space. When electrical problems in the capsule required him to take manual control of the craft toward the end of the flight, he brought it back to Earth using dead reckoning and good old cowboy stick-and-rudder skills. It was the perfect blend of technological accomplishment and aw-shucks American badassery. Suddenly NASA was on a roll.

The Gemini project was next. Less glamorous but intensely practical, these ten flights were crewed by two astronauts each, hence the name—denoting the "Twins," Castor and Pollux, in Greek mythology. The Gemini missions went up from 1965 through 1966 and focused on achieving some of the operational tasks, like tracking, approaching, and docking with another spacecraft, that missions to the moon would have to accomplish. It was on Gemini 4 that Ed White became the first American spacewalker in 1965. On Gemini 8, Neil Armstrong successfully halted a capsule spin that threatened to kill him and his crewmate, Dave Scott, furthering Armstrong's reputation as a man whose blood turned to radiator coolant under pressure. The final Gemini flight, Gemini 12, took place in November 1966 and featured extravehicular activities (EVAs, or spacewalks) by the hyperfocused rookie Buzz Aldrin, who took photographs, brought in a micrometeorite collection experiment attached to the exterior of the capsule, and generally demonstrated that astronauts could work outside a spacecraft in microgravity. Gemini 12 was commanded by the genial Jim Lovell, who would later fly on the Apollo 8 and 13 missions, both of which flights were miraculous, though for different reasons.

But NASA was just warming up for the main event. In May 1961 President Kennedy had challenged the nation to send a man to the moon and bring him safely back home before the end of the decade. With Mercury and Gemini concluded, NASA turned its attention to Apollo, which aimed for a lunar landing. The agency was rapidly expanding and, in the process, soaking up as much money as Congress could throw at it. And with an ambitious schedule of Apollo flights, each of which would require a crew of three, NASA needed more astronauts.

The Navy wanted naval officers involved in the space program, just as the Air Force wanted its pilots to be represented. In fact, it was the Navy that identified Bruce McCandless as a candidate for the fifth astronaut selection and invited him to apply. He didn't need much coaxing. The basic qualifications for selection were clear enough: The successful applicant would be no older than thirty-six, under six feet tall, hold a bachelor's degree in engineering or the biological or physical sciences, and have test-pilot experience or at least 1,000 hours flying time in jet aircraft. Simple enough, right? Cut and dried. Except that it wasn't. As astronauts Mike Collins, Mike Mullane, and Leland Melvin have all written, there was extensive physical, psychological, and even social screening involved in the selection process. Along with the intrusive physical tests described by Tom Wolfe and a number of NASA veterans, applicants were also assessed on how they looked, talked, and interacted with others. In a recorded interview, Dad said he duly complied with the dictates of his naval superiors and the agency's application protocols but didn't seriously expect to be selected on the first try. He was too young, he thought. He'd have a better chance once he earned his PhD.

NASA received thousands of applications and inquiries from would-be astronauts. From this number, 351 candidates were deemed to be qualified, at least nominally, and ultimately nineteen men were chosen. Lantern-jawed Mississippian Fred Haise, possibly the nicest guy in the program, received the group's top score. And Dad was wrong about his application. In April 1966, at the age of twenty-eight, he became the youngest member of Group 5, joining a class that included, among others, Vance Brand, Charlie Duke, and Jack Swigert. SPACE TEAM GETS 19 NEW ASTRONAUTS, read the headline of one UPI story. I'M OFF TO MOON, DEAR, said another, this one about my dad's selection in particular. California's *Chico Enterprise* chimed in with NEW ASTRONAUT GOT SPACE BUG WHEN HE WAS 10. The Group 5 selection was the largest to date, and brought the number of astronauts to a record-high fifty. Veteran John Young dubbed the new group the "Original Nineteen" in a sardonic reference to the Original Seven who had pioneered the job. Other observers were more biting, calling the group the "Excess Nineteen" to express a belief that NASA now had what seemed like a surplus of space travelers.

ACCORDING TO SPACE HISTORIAN Matthew Hersch, Bruce McCandless and his Group 5 colleague Don Lind were effectively treated as "scientist-astronauts," akin to those selected in the fourth and sixth astronaut groups, an implicit reflection of their shared lack of the test-pilot experience valued by Deke Slayton and other NASA managers at the time. This categorization might seem strange to younger NASA enthusiasts, given that so many current astronauts are, in fact, scientists. In the sixties, though, it was an important distinction—and not, for McCandless and Lind, a helpful one.

Astronaut Brian O'Leary made the difference clear after he resigned from the program in 1968. A member of the Group 6 scientist-astronaut class, O'Leary wrote that he and several of his comrades were flat-out told by Slayton that NASA would have no use for them in the foreseeable future. O'Leary was an odd choice for the astronaut corps from the beginning. He'd earned his PhD from Berkeley in 1967 with a dissertation about possible properties of the surface of Mars. Even at the time of his selection, he was disenchanted with the American military's role in Vietnam. He hated Houston, which he described as a place of "heat, humidity, smog, gray skies, rain, flat terrain, mud and general dullness." Once inducted, he complained bitterly about what he saw as the anti-scientist prejudice in the program, describing the astronaut corps as "fifty clean-cut, erect, alert" military types with little scientific interest or imagination. He wrote that forcing astronomers and physicians like those in his class to learn to fly jet airplanes was a waste of time and, potentially, of lives.

In fairness to O'Leary, four astronauts had been killed in T-38 accidents in the previous three years. Further, cosmonaut Yuri Gagarin, widely celebrated as the first man in space, also died, under somewhat mysterious circumstances, in a jet airplane crash in 1968. O'Leary may have been right about the piloting requirement, but he misjudged the individuality of his comrades. Within a few years, Rusty Schweickart would be studying transcendental meditation; by 1979, as a science advisor to California Governor Jerry Brown, he was "sitting in a Pasadena auditorium with a metallic star pasted on his forehead as dancers circled him, chanting for the elimination of nuclear power plants." Donn Eisele took a position with the Peace Corps after leaving NASA. Mike Collins departed Houston to join the U.S. State Department and subsequently ran the Smithsonian's new Air and Space Museum. Jim Irwin became an evangelist. Ed Mitchell conducted experiments in extrasensory perception (ESP) while in space. He had a sort of epiphany as he gazed at Earth on his return from the moon on Apollo 14 and later founded the Institute of Noetic Sciences to explore the ability of some individuals, as an institute newsletter put it, to "perceive information not presented to any known sense and blocked from ordinary perception." The astronauts weren't really automatons, devoid of imagination or eccentricity. They just had to act like they were.

The perception that Bruce McCandless was more scientist than pilot, says Hersch, ultimately delayed his progression in the flight rotation. In fact, my father defined himself as neither scientist nor pilot but something in between and quite possibly better: an *engineer*. He had little patience for theoretical science. He shook his head when asked if he might enter academia after his career in the space program. By the same token, he saw flying as a means rather than an end in itself, though he enjoyed it. Where he excelled was in the application of technology to real-world problems. He liked fixing things. In fact, he liked fixing things the way Captain Ahab liked chasing whales. It didn't matter what. When the drain line from our house broke one Thanksgiving, we went out in the rain with shovels to dig up the pipe, locate the problem, and replace the section of PVC. When the heat and humidity of Houston finally destroyed our roof, we put on a new one ourselves. He usually let me have the tape measure to make cuts on the plywood decking, but he always made me use the Pythagorean Theorem to figure the length of the hypotenuse when we were doing corners. In all the years I lived with my parents, I never saw a single repairman enter our home. Bruce McCandless replaced broken windows, rebuilt three carburetors, got the dishwasher running again, and serviced the brakes on the Suzuki 100 I rode back and forth to high school. Dad did everything himself. It was a point of honor. Ron Sheffield, a Martin Marietta manager who worked with him, recalled that Bruce McCandless was the only man he ever met who not only could, but actually did, disassemble and rebuild his own microwave oven. My father was a man with an abiding love of socket wrenches and chain saws, air compressors and voltage meters. In those days the Snap-On Tools van periodically visited our neighborhood. If Dad got wind of the visit, he would jump in his car and chase the vehicle through the streets like a kid on the trail of an ice cream truck. His practical, hands-on mechanical know-how would eventually come in handy. In the short term, though, Bruce McCandless was an odd duck, stuck between the cool kids of the test-pilot fraternity who were happily dancing in the gym and the geeky geologists and astrophysicists who stood outside, barely able to hear the music.

Because my father had tunneled through the required course work for his doctorate, all that remained (*all!*) was his dissertation. He wasn't thrilled about leaving Stanford before he could finish his studies, but the opportunity to join the astronaut program was something an ambitious young naval officer with a lifelong dream of walking on the moon simply couldn't pass up. Alan Shepard called our little house in Mountain View one day in April 1966 and asked to speak with Bruce.

"He's not here at the moment," Mom said. "I think he's out flying."

"Okay, could you—"

"Flying, or over at the campus, doing one of his experiments."

"Experiments."

"Oh, yes. In plasma physics. Can I give him a message?"

If Bruce McCandless hadn't cinched the job before the call, he certainly had it afterward. Dad dialed up Shepard as soon as he got home. He listened carefully to the famous astronaut's words and paused for a moment to consider. But only for a moment. Then he grinned and said yes, yes of course, and our new lives began. It was a great moment for all involved.

Except, perhaps, Oscar Buneman. The professor showed up at a press conference convened on April 4, 1966 to announce Bruce McCandless's selection as an astronaut. The conference took longer than expected. My father, it seemed, was late for class. Buneman voiced his dismay that his star pupil was even thinking about interrupting his doctoral work. "Don't treat him as a pilot," he told the assembled journalists. "He's an academician!"

As with much of his work, Professor Buneman's words in this regard proved to be prophetic.

6

YOU'LL GET USED
TO THE HEAT

Texas was as likely a place for a high-tech spaceflight facility as the nation's capital was for a cattle drive. Nevertheless, Houston was chosen as the location of the Manned Spacecraft Center in 1961, beating out such rivals as San Diego, Shreveport, and Tampa. The winning site wasn't actually in Houston proper, but twenty-five miles southeast, on pastureland owned by Rice University on the shore of Clear Lake. Some observers questioned the choice of location. The lake, as many have pointed out, is hardly clear. The land is scrubby and unremarkable. And the region, like the rest of the Gulf Coast, is subject to hurricanes, such as Hurricane Carla, which came ashore just a few days before the site award was announced. The storm ripped up trees, flooded neighborhoods, and killed thirty-six people. A young reporter named Dan Rather made a name for himself during the disaster by reporting live from nearby Texas City as cyclonic winds swirled around him, thereby changing the way hurricanes, tropical storms, and mall openings would be covered by television journalists for the next sixty years.

But there were advantages to Clear Lake, too. Ellington Air Force Base, vital bulwark in America's air defenses against Mexico, stood just to the north, with long landing strips and room to grow. There were two fine universities nearby, Rice and the University of Texas, and another, Texas A&M, that could only get better. The site was accessible by barge traffic, which was important because NASA envisioned transporting some of its space machinery by sea. It was near a large metropolitan center, and the surrounding land was cheap, since it wasn't good for much other than grazing. Smelling money, the future, and historical relevance, Houstonians lobbied whoever would listen. NOT ALL OUR SNAKES ARE POISONOUS, they promised. And YOU'LL GET USED TO THE HEAT. Vice President Lyndon B. Johnson, Speaker of the House Sam Rayburn, and Congressman Albert Thomas were all powerful men in Washington. They were also Texans, and they wanted Houston. So Clear Lake got the nod.

In many ways the marriage of Houston and NASA was a union of opposites. Houston was a city built on mud. It started with the Allen brothers' first warehouse on the banks of Buffalo Bayou and continued through the dredging of the bayou and parts of Galveston Bay to create the Houston Ship Channel and the frantic industrial vampirism of the oil booms that fueled the city's growth. Houston prospered by looking down, digging in, getting dirty. Unlike the staid mercantilism of Dallas or the more intellectual preoccupations of Austin, Houston had a muscular roughneck vibe. NASA by contrast was all about aiming upward. Its world was white shirts and slide rules, clean rooms and IBM punch cards. Grafting the Manned Spacecraft Center onto the city was like mounting a diamond on a ring made of rebar. But somehow it worked. Construction began in 1962, and the center was officially declared open in June 1964. Clear Lake City, a planned community, sprang up alongside it, a sprawling suburb of ranch-style houses with tidy, treeless lawns, a rec center, two swimming pools, and a golf course. And though Houston was eager to add the diamond to the ring, space center employees didn't necessarily feel the same way about becoming part of the metropolis. When the city announced its intention to annex Clear Lake in the mid-seventies, a protracted battle broke out. Area residents took to the courts, determined to keep their schools white and their property taxes low. FREE THE CLEAR LAKE 25,000, said the bumper stickers. It didn't work. Houston's annexation bid survived the legal challenges, and the marriage was eventually made official after years of common-law association. By then the conflict was irrelevant anyway. In the mind of the public the connection had been cemented when Neil Armstrong radioed back from the moon: "Houston, this is Tranquility Base. The Eagle has landed."

WHEN BRUCE MCCANDLESS JOINED NASA in 1966, we traveled from northern California to the Gulf Coast, where we moved into a 2,400-square-foot, two-story brick and cedar-sided house in El Lago, three miles from the Manned Spacecraft Center (which was renamed the Johnson Space Center in 1973). El Lago is a bedroom community on the eastern shore of a brackish body of water called Taylor Lake. There were 750 residents when we arrived. By 1990, fueled by the expansion of NASA, the population had swelled to more than 3,000. Growing up, I usually said I was from Houston, unless I was talking to someone *else* who was from Houston, in which case I said *Seabrook*, which was a fishing town on Galveston Bay and the location of the nearest post office. Technically I lived in El Lago, which had its own mayor and cops. But since El Lago was more a philosophy of lawn care than an actual city, Seabrook was probably the best answer. I never said Clear Lake. Clear Lake wasn't a place, it was a plat. Clear Lake was an advertisement. It would suck the soul clean out of your body.

My first NASA memory is of hearing the news on the car radio about the Apollo 1 fire and the deaths of Gus Grissom, Ed White, and Roger Chaffee. It was January 27, 1967. In my mind, the tragedy seems to take place on the same night my parents stopped on the side of NASA Road One to rescue a stray dog—a basset hound, I think. It's not so much the news itself I recall as the reaction it elicited from my mom and dad, who discussed the event in the front seat of the Jeep in hushed tones. The crew had been practicing for the launch of Apollo 1 in Florida, fully suited and deployed in the command module as they would be at liftoff, which was scheduled to take place just a few days later. The atmosphere in the cockpit was pure oxygen. When a short in the wiring sparked, the oxygen fed flames that flashed through the command module's nylon netting and foam pads. The crew tried to get out, but the capsule's hatch was designed to open inward. A rapid exit was difficult under any circumstances, but it became practically impossible with the sudden rise in air pressure in the capsule as the fire intensified.

Gus Grissom, the second American to fly in space, was opinionated and irascible but a gifted engineer. Ed White was brilliant and beautiful, perhaps the best of NASA's second generation of spacefarers. His space-walk in 1965 had vaulted him to a place beside the immortals, Al Shepard and John Glenn. Fresh-faced Roger Chaffee was the rookie on the crew, inexperienced but bright and well-liked. NASA controlled what the public heard about the tragedy, as it would with subsequent disasters. The official line was that Grissom, White, and Chaffee perished instantly. Insiders knew this was a lie. The men cried for help as the flames flashed through the cabin. It was an unspeakable event—some thought insurmountable.

The Apollo 1 fire was the agency's original sin, a tragedy that cast a long, sobering shadow over the men who were waiting to fly. It took him many years, but my father later admitted what had immediately become clear. The astronauts were too optimistic. In their eagerness to ride a rocket, see the solar system, and beat the Soviets, they were willing to forego a lot of hard questions and trust their lives to an army of administrators, engineers, and contractors. Bruce McCandless was just as guilty as the rest. When he was interviewed in 1966 for a spot in the astronaut corps, he was asked if he'd be willing to travel to the moon and back with a fuel reserve of just *2 percent*. "I would," he said at the time. Years later, he winced at that cavalier response. Just as many of us can remember where we were when we saw *Challenger* break apart in January 1986, my parents always remembered the Apollo 1 fire. The dangers didn't stop my father. They didn't stop any of the men waiting to fly. But it did remind them of the hard fact that space travel could kill them, and that they were going to have to take a more active role in the design and operability of the machines they were going to rely on.

Eventually the pall faded, and the program started up again, this time with myriad improvements, large and small, to the spacecraft. The command module was redesigned. The oxygen content in the capsule was significantly reduced while the vessel remained on Earth, and the escape hatch was simplified and modified to open outward, so that the increased cabin pressure generated in the event of a fire wouldn't impede egress. In the meantime, those involved shared the mounting excitement of working on Apollo, the greatest engineering project in history.

The astronauts trained for all sorts of contingencies. In June 1967 many of them flew to Iceland for a hands-on course in geology—central Iceland's desolate terrain being judged the closest available stand-in for conditions on the moon. And because NASA feared a space capsule might not return to Earth exactly where it was supposed to, the astronauts practiced coping with the challenges presented by a variety of extreme environments. In August 1967, for example, Bruce McCandless traveled to Washington State for an Air Force Survival School training session focused on desert conditions. Fieldwork took place in an area northeast of Pasco, Washington, where a schedule said, "The astronauts as a group will learn to make emergency clothing from the parachutes which lower their space vehicle to earth. Their knowledge of navigation will be reviewed. Emergency ground-to-air signals will be constructed and later evaluated by School personnel flying above the training site." My dad also went on a survival trip to the jungles of Panama later that year. He and his colleagues rafted the Chagres River, ate boa constrictors, iguanas, and armadillos, and learned to catch rainwater in the fronds of the local vegetation.

As tough as it was to get into the astronaut corps in the first place, the jockeying for position among the group never really let up. Indeed, the very first entry in Bruce McCandless's 1967 diary is a reminder to turn in his numerical rating, in a sealed envelope, of his nineteen Group 5 peers for review by Deke Slayton and Al Shepard. At least initially, the competition was part of the excitement. My dad's early years in the astronaut program were euphoric. It must have felt like being back at the Naval Academy, with added overlays of celebrity, heroism, and technical innovation he was expected not only to enjoy but also to create. Just as at the Academy, he was the youngest kid in his class. A twenty-eight-year-old rookie, he suddenly found himself sharing office space in the Manned Spacecraft Center's Building 4 with newly minted American legends like Deke Slayton, Gordon Cooper, and happy-go-lucky Wally Schirra. Dad studied orbital mechanics, geology, and atmospheric physics with men who were just as smart and dedicated as he was. He flew government-owned jet aircraft, the fabled T-38 Talons, sleek and swift, to out-of-town PR engagements and training sessions.

While piloting a T-38 demanded skill and concentration, it had the immense advantage that no one was shooting at you as you passed. In October 1967, while on a bombing run over Hanoi, John McCain's A-4 Skyhawk was hit by a surface-to-air missile. The crippled plane went into a spiral. McCain ejected, breaking both arms and a knee in the process, and parachuted to a landing in a shallow lake. He was captured and beaten by North Vietnamese civilians and imprisoned in the North's "Hanoi Hilton" with other captured American servicemen. The dashing, easygoing McCain would spend the next five years as a prisoner of war, subjected to solitary confinement and to torture so severe that for the rest of his life he was unable to raise his arms over his head.

Meanwhile, in Houston, bankers and builders wanted Bruce McCand-less's business. A Chevrolet dealer in La Porte offered him the GM product of his choice, available on a $1-per-year lease. (He chose, successively, a Corvette, a Chevelle SS, and the big Suburban we eventually bought.) Politicians and movie stars angled to meet him. He got fan mail, though he'd have been the first to admit he hadn't really *done* anything yet. He worked and traveled with Alan Shepard, and some of his favorite stories involved the man who was clearly the astronaut office's first among equals. After a long day of collecting rock samples in the deserts of New Mexico, for example, the astronauts returned to El Paso to spend the night in a local hotel. A crowd of admirers was waiting for them when they got off their bus. An elderly woman was first to approach, saying excitedly that she wanted to get the famous Al Shepard's autograph. Shepard said, "Ma'am, you're in luck. He's right here." He put his hand on the small of my dad's back and thrust him into the maw of the gathering crowd. Dad then signed "Alan D. Shepard" autographs for almost an hour after Shepard himself disappeared into the air conditioning.

It was heady stuff. I have a photograph of Bruce McCandless taken in 1969. In the shot, he's standing with Jim Irwin and Charlie Duke outside a local Ramada Inn. The picture is unposed, as if taken surreptitiously by some pioneer paparazzo. Like the other men, my father is lean and loose-limbed, vaguely vulpine, an advertisement for adrenaline and aviator glasses. Bruce McCandless was a fighter pilot and an astronaut, brilliant and dashing. He wanted the world, and the world wanted him back. In that frenetic era, at a time when the nation seemed to be coming apart, torn by the hatreds engendered by Vietnam and racial inequality, the astronauts were Establishment eye candy, untouchable and unafraid. They were the president's pen pals, Tom Swift's bachelor uncles, one half Boy Scout and two-thirds matinee idol. They knew Science but didn't bring her to parties. They could fly anything. They cut their hair short and at precise geometric angles to minimize drag. It was easier to slip into the future that way.

It was a year of airplane hijackings and student protests. Richard Nixon ordered the secret bombing of Cambodia, and almost 12,000 American servicemen died in Southeast Asia, fighting an elusive enemy in the service of an even more elusive cause. In Houston, though, it was hard to see beyond the bubble of adulation and energy building around Apollo, a project that combined massive amounts of human intelligence and labor with godlike ambition and huge infusions of patriotic fervor. The will was there. The brainpower was there. Incredibly, even the money was there. For two years during the Gemini and Apollo programs, space-related expenditures spiked to more than 4 percent of the annual federal budget. Hundreds of thousands of Americans worked on the project. Millions more watched with fascination and growing excitement. Even the novelist Norman Mailer, exhaustingly agnostic, skeptical of the technological hubris that led to the launch, was nevertheless unsettled by Apollo's scope and audacity. Visiting the Vehicle Assembly Building at Cape Kennedy, where the Saturn V rockets were put together, he wrote of his alter ego, "Aquarius":

The change was mightier than he had counted on. The full brawn of the rocket came over him in this cavernous womb of an immensity, this giant cathedral of a machine designed to put together another machine which would voyage through space. Yes, this emergence of a ship to travel the ether was no event he could measure by any philosophy he had been able to put together in his brain.

After the Apollo 1 fire, Americans didn't venture into space for almost two years. A series of unmanned test flights paved the way for Wally Schirra, Donn Eisele, and Walt Cunningham's successful Apollo 7 mission in October 1968. NASA then made the audacious, possibly ill-advised decision to send the crew of Apollo 8 to the moon. Against long odds, Frank Borman, Bill Anders, and Jim Lovell made an almost perfect flight, demonstrating conclusively that travel to Earth's satellite and back was well within the country's technological abilities. Apollo 9 and 10 were technically demanding but less dramatic rehearsals for the big event. Apollo 11 would attempt what had long been thought impossible.

Step-by-step, mission by mission, the moon grew closer. Like any other kid I ate my frosted Pop-Tarts and tried to figure out what I was seeing. I'm sure the images in my memory come from not one but several different missions. It's early on a weekday evening. Dad's still at work, and the TV's on. A reporter is talking over a picture of a rocket standing in the distance beside a giant metal tower. The scene shifts. Three astronauts are dispensed from an unremarkable boxlike structure in their perfect white suits, adorned with nozzles and emblems, hermetically sealed to avoid contamination from the rest of us sticky, mouth-breathing mortals. They wear yellow galoshes. Their heads inhabit fishbowl helmets like they've gotten stuck in there somehow, but otherwise the figures seem capable, eager, jolly almost. Like good, solid suburban men everywhere, they carry briefcases, but here the cases are actually attached to their suits because the contents are so important: tuna-fish sandwiches, the crossword puzzle, a love letter to God, a map of the universe just in case they get lost. The astronauts move with an odd, stiff-legged gait. They're accompanied by men in white jumpsuits and hardhats—wrench-bearing priests in this church of glamorous science. Flashbulbs pop and reflect off the waiting transfer van like tiny lightning strikes. In their pressure suits, the space travelers are slightly larger than those around them, which seems fitting. They grow bigger as we watch. They're giants. They're legends. They're going to glory, riding an elevator up to the nose of the massive Saturn V not only to take control of the snorting propulsive technology below them but also to touch the mythology above. They live in 3-bedroom ranch-style houses on modest lots, but soon they will enter the heavens to become the gods themselves—Mercury, the Twins, fiery Apollo. We learned about Apollo in school. Every morning he takes the reins of a golden chariot and drives his divine horses across the sky, so brilliant and beautiful a sight that we call it the sun. This man is not the sun. This man, says the reporter, is John Young. He waves a final time and steps into the van. He will be sealed into the rocket like a pharaoh in a pyramid. We hear the countdown. The rocket climbs out of a nest of fire and smoke and disappears into our bookshelves. A successful launch: I part the curtains behind our sofa to stare out at the sky. It hasn't changed. I'm reverent, but not enlightened. Next comes a commercial for Rice-a-Roni, and the world deflates again.

ALONG WITH OTHER TECHNICALLY minded astronauts, including scientist-astronauts of the 1965 and 1967 groups, Bruce McCandless started his NASA career in the Apollo Applications program. As the name suggests, Apollo Applications was established to develop ways to use Apollo hardware for an ambitious range of space-exploration projects—a permanent space station, a mission to Mars—following the moon landings. In his book *Carrying the Fire*, Michael Collins describes one of these projects: evaluating competing designs for a method by which an astronaut could leave the safety of his vessel and move about, untethered, in space.

One proposed device was a set of "rocket boots," featuring propulsive systems built at first into the astronaut's footwear and then later incorporated in a sort of sled that was merely controlled by the user's feet. Bruce McCandless tested what he called the "jet shoes" in the summer of 1969. In a memo to Deke Slayton, he concluded that they were impractical and unworthy of further investment. While Slayton concurred, rocket boots nevertheless remained a contender for several more years. Another alternative for extravehicular locomotion was a gas-firing pistol called the "hand-held maneuvering unit." Ed White used an oxygen-fueled version of the device on his spacewalk in 1965. Collins tested it himself, with mixed results, on Gemini 10. Developed by NASA engineers, the pistol was small, light, and easy to operate. The main problem with the device was aligning the unit's propulsive force with the astronaut's center of mass. Even a slight miscalculation in this regard could send the astronaut spinning off target. My father once commented that the pistol was elegant in concept but fiendishly difficult to operate, and that he spent twelve years trying to kill it.

The more promising option, in Bruce McCandless's opinion, was something called the Astronaut Maneuvering Unit, or AMU. Basically a gas-propelled jetpack, the device was seen as exotic even by the men who were going to test it. Mike Collins called it "a little spacecraft in itself" that was "really far out." The AMU was originally developed by the Air Force for possible use on its planned Manned Orbiting Laboratory, a military space-station project that never came to fruition. The AMU had twelve thruster nozzles that used nitrogen gas to "positively expel" hydrogen peroxide via a bladder in the hydrogen peroxide tank. The heat generated by the H_2O_2 necessitated development of a pair of metallic coveralls ("stainless steel pants," my dad called them) to protect the astronaut's lower extremities from the exhaust. The device consisted of a pressure suit, a chest-pack life-support system, and a backpack maneuvering unit. Together, the components weighed 168 pounds. The AMU's thrusters, which were fired in pairs, produced 2.3 pounds of nominal thrust and were divided into a primary and an alternate system. Importantly, the machine had what engineers call six-degrees-of-freedom maneuvering capability—that is to say, its thrusters could be controlled to provide movement (1) up and down; (2) forward and back; (3) left and right; and to create (4) pitch, (5) yaw, and (6) roll. Basically, six-degrees-of-freedom maneuvering capability means that your machine can move any direction you want it to.

In June 1966, shortly after Bruce McCandless joined the space program, Gene Cernan exited the capsule of Gemini 9, 150 miles above the Earth, to perform the first in-flight test of the AMU. His initial task was simply to get to the device, which was mounted on the spacecraft aft of the capsule. The attempt did not go well. Cernan, a lean, athletic man, was hindered by the stiffness of his pressurized, heavily insulated space suit, including the stainless steel pants, and by the fact that the Gemini spacecraft had only two handholds. With little to anchor himself to, he continually found himself demonstrating Newton's Third Law of Motion—namely, that every action creates an equal and opposite reaction. Cernan overexerted himself to the point where condensation created on the inside of his helmet visor by his own breath and body heat rendered him essentially blind. His description of the EVA occasionally sounds like something H. P. Lovecraft could have written. In the darkness of space, Cernan's umbilical tether became a "snake," a "worm," and an "octopus" that left him "slipping in puddles of space oil, with no control over the direction, position, or movement of my body" as "the naked sun, an intense ball of gleaming white fire, stared at me, a tiny interloper in its realm." As a result of Cernan's exhaustion, the AMU test was called off not long after he managed to get himself situated in the unit. The failed exercise almost ended badly. NASA flight controllers worried that Cernan might not even make it back inside the capsule. It turned out that the astronaut lost thirteen pounds of sweat on his brief sojourn outside the spacecraft.

Astronauts Mike Collins and Dick Gordon had similar difficulties with space walking on Gemini flights 10 and 11, respectively. Both men were tethered to their Gemini capsule. Collins used a handheld maneuvering unit that provided some assistance and was able to complete his assigned tasks. Gordon's assignment was to attach a tether to the Agena space vessel his Gemini capsule had docked with, for purposes of an attitude stabilization experiment. He found working in space to be so tiring that at one point he straddled the nose of the Agena space vehicle like Slim Pickens riding the bomb in *Dr. Strangelove*. He did manage to attach the tether. But because Gordon was sweating so much that he couldn't see straight, Commander Pete Conrad ordered him to re-enter the capsule. In light of these problems, a second planned attempt to test the AMU on Gemini 12 was canceled. NASA administrators seemed to develop a dislike not only for the AMU but also for spacewalks in general.

This was the backdrop for Bruce McCandless's assignment to the jetpack project as a rookie. As intriguing as the possibilities were to the young engineer, the AMU was clearly a sort of developmental backwater for NASA, especially as there were no plans for use of jetpack-powered EVAs during the Apollo program. Nevertheless, Dad plunged into the work the only way he knew how—completely. He made his first simulated flight of what was now called the MMU, the Manned Maneuvering Unit, on December 15, 1966. A photograph of him in *The Sacramento Bee* in December 1968, two years later, shows him being held by what looks like a giant robotic claw, operating the space jetpack in the zero-G simulator at the Martin Marietta plant in Denver. Operating machinery while in the clutches of a giant claw is no longer permissible under various OSHA regulations, of course, but the developers of the device eventually found other ways to simulate operation of the jetpack, including neutral-buoyancy testing in vast swimming pool–like water tanks. The young astronaut would go on to log hundreds of hours over the next two decades in various sorts of simulated MMU operations.

While Apollo Applications work was intriguing, Bruce McCandless soon realized he had to be closer to the action if he ever wanted to fly. Thus, the biggest plum of his early career was his assignment to be capsule communicator, or capcom, on flight director Cliff Charlesworth's green team for the Apollo 11 mission. The capcom was designated to be the single voice the astronauts in space would hear and respond to in Houston. As such, anyone in the position had to have an intimate knowledge of the mission plan, the myriad technical systems on board the spacecraft and how they worked, and the men actually up there in charge of the machines. From the very first Mercury mission, it was agreed that only a fellow astronaut would do.

No one joins the astronaut program to be a capcom. It's like being the backup quarterback who relays plays from the sidelines; you're part of the action, but no one's going to remember you after the game. Still, it was a coveted assignment, and a sign that bigger things were on the horizon. Indeed, the other two astronauts who did capcom duty for Apollo 11, Charlie Duke and Ron Evans, both flew on later Apollo missions.

Transcripts of the communications between Bruce McCandless and the Apollo 11 astronauts—Armstrong, Aldrin, and Collins—are online, and portions of the audio recordings are featured in various film and television documentaries. The conversations range from highly technical (*"For today, we'd like you on P23 to make a trunnion bias determination . . . The bias that you get beforehand should be incorporated, that is a PROCEED on NOUN 87 after you get two consecutive measurements equal to within 0.003 degrees."*) to frankly frivolous. Each morning, the capcom would read newspaper stories to the crew. This was not much of a task for Bruce McCandless, actually, since newspaper reading was one of his favorite pastimes. On July 19, 1969, he reported that the Soviets' unmanned lunar mission, Luna 15, which was scheduled to land on the moon just before Apollo 11, had ceased transmitting radio signals and was evidently lost. Other big stories: British parliamentarians were asking for assurances that a midget submarine to be launched in Loch Ness would not harm any monsters living in the long lake's inky depths, and Mexican authorities announced that American hippies would no longer be granted visas to travel south of the border until the counter-culture's heroes cut their hair and took a bath.

Charlie Duke handled the capcom duties as the Apollo 11 astronauts descended to and landed on the moon—a process that took place, oddly enough, with Armstrong and Aldrin standing up in the lunar module, as there were no seats in the craft. After the Eagle touched down on July 20, the plan was for the astronauts to rest before they exited the spacecraft to explore the surface. My dad was next up as capcom, and *he* planned to use the scheduled siesta as a dinner break. But the astronauts were too keyed up to rest. Dad was on the way home at the time the decision was made to proceed with the moonwalks. He pulled into our driveway only to be met by my mother, who ran out of the house as if it was on fire. Mom took her space duties seriously. She'd just gotten a call from Mission Control, and had a message to relay. "Go back!" she yelled. "The astronauts can't sleep! They want to go out now!"

Dad reversed, turned his Jeep around, and screeched out of the cul-de-sac in front of our house for the ten-minute drive back to the space center. The moon, a waxing crescent, was standing thirty degrees above the western horizon, and my father slipped into a sort of reverie as he drove toward it on NASA Road One. He turned right at the U-Tote-Em where I bought my baseball cards. He skirted the north shore of Clear Lake, passing the old West mansion as he neared the space center. And still the moon floated serene and imperturbable in front of him like a black-and-white photograph of itself, Earth's gravitational remora, her pale silent sister, movie star and legend, goddess and mirage. Bruce McCandless had just turned thirty-two. He was only a few months older than his father had been when he faced down the giant Japanese battleship *Hiei* in the midnight seas off Guadalcanal. He was a sworn son of science, an engineer, a distant nephew of Sir Isaac Newton. He knew the formulas required for achieving orbital velocity, could tell you the fuel mixtures you needed, the stages and timing of rocket-booster separations. He brushed sentiment away like so many spider webs. But even he was having trouble today believing that *human beings*—his colleagues and friends—were up there in the sky, getting ready to do something no one had ever done before. He was going to be part of it. He would be talking to two men as they walked on the moon. The warm Gulf air blew through the open windows of the Jeep without offering much relief from the heat. Sweat beaded on the small of his back. His pulse danced in his wrists. Dad hadn't quite reached his lifelong goal of touching the lunar surface, but he was close. He was almost there.

He could feel it.

So IT WAS THAT my father was on the job, talking to Neil Armstrong and Buzz Aldrin, when Armstrong eased himself down the ladder of the Eagle to set foot on the moon. We take it for granted now. At the time, though, it was unclear whether an astronaut could even *stand* on the moon. How deep did the dust go? Were volcanic lava tubes lurking beneath the gritty surface of this lunar hellscape, just waiting for some unlucky earthling to fall into? Some 600 million people around the world were watching as Neil Armstrong prepared to find out. It seemed to take an eternity for him to get those last few feet down the ladder. NASA's transcript of moon-to-ground communications that day captures the mix of mundane and magnificent:

> 109:23:39 Armstrong: I'm at the foot of the ladder. The footpads are only depressed in the surface about 1 or 2 inches, although the surface appears to be very, very fine grained, as you get close to it. It's almost like a powder. Ground mass is very fine.
>
> 109:24:14 Armstrong: Okay, I'm going to step off the LM now.
>
> 109:24:26 Armstrong: THAT'S ONE SMALL STEP FOR MAN, ONE GIANT LEAP FOR MANKIND.
>
> 109:24:49 Armstrong: And the . . . the surface is fine and powdery. I can pick it up loosely with my toe. It does adhere in fine layers like powdered charcoal to the sole and sides of my boots. I only go in a small fraction of an inch, maybe an eighth of an inch, but I can see the footprints of my boots and the treads in the fine, sandy particles.

109:25:31 McCandless: Neil, this is Houston. We're
copying.

109:25:49 Armstrong: There seems to be no difficulty
in moving around as we suspected. It's even perhaps
easier than the simulations at one-sixth g that we
performed in the various simulations on the ground.
It's actually no trouble to walk around. Okay. The
descent engine did not leave a crater of any size.
It has about 1 foot clearance on the ground. We're
essentially on a very level place here. I can see some
evidence of rays emanating from the descent engine,
but a very insignificant amount.

109:26:55 Armstrong: Okay, Buzz, we ready to bring
down the camera?

109:27:00 Aldrin: I'm all ready. I think it's been all
squared away and in good shape.

Armstrong may have bungled the delivery of his triumphal pronounce-
ment. It was meant to be one small step for *a man*, not for man generally,
since for man generally it was a *giant leap*. No matter. Everyone knew what
he meant. It was one of the great and truly magical moments of human history, and Bruce McCandless knew enough to keep quiet and let it unfold.

The first excursion of men on the moon lasted just over two hours. Neil
and Buzz, as everyone now referred to the space explorers, gathered rock
and soil samples, tested how best to walk on the lunar surface, took photographs, and planted the American flag. Every minute was scheduled on
a checklist. The astronauts did the best they could, given the constraints of
their pressure suits and the sheer surrealistic wonder of being present in a
place that had existed for millennia as a kingdom of dreams.

Bruce McCandless was fascinated by the grainy black-and-white video
stream coming from space. At some point, he couldn't remember when, the

head of the astronaut office came and sat beside him. It was his boss, Deke Slayton, World War II combat veteran, former test pilot, and member of the Original Seven. He was a rawboned, imposing individual with thick eyebrows and a gruff, drop-the-bombs-now sort of voice. He was also, as everyone knew, no one to trifle with. The minutes wore on. The grizzled veteran cracked his knuckles and leaned closer.

"Bring 'em in," he murmured.

My dad's heart froze. He glanced over at Slayton. The older man's face was impassive, but he turned to look at the young astronaut full on. The arrowhead of his widow's peak was aimed straight at my father.

"Bring 'em in," he said. "They're getting tired."

Bruce McCandless gulped. This wasn't on the checklist, which was a sort of Bible that everyone associated with the flight knew in intimate detail. It was the checklist that dictated what the two moonwalkers would do and when they would do it. There was still time on the clock and a number of tasks to be completed. The astronauts were obviously enjoying themselves. And anyway, he, Bruce McCandless, didn't have the authority to bring *anyone* in. The flight director is to a mission what a captain is to a ship—and Dad wasn't about to pipe up and ask Flight Director Cliff Charlesworth if he, Bruce McCandless, could direct Neil and Buzz to go back to the lunar module without a damned good reason.

The young astronaut shook his head. Slayton looked on for a few moments, the irritation plain on his face. Then he got up and walked away.

THE PHILOSOPHER ERIC HOFFER called Apollo 11 the "triumph of the squares," as if it were some sort of domestic cultural statement, performance art by the pocket-protector set. He underestimated the diversity of the men and women who made it happen, and the awe of the generations since who have adopted the achievement as their own. Wernher von Braun, when asked how he evaluated the significance of the act of putting a man on the moon, answered, "I think it is equal in importance to that moment in evolution when aquatic life came crawling up on the land." Essayist Julie Wittes Schlack, who watched television coverage of the Apollo 11 landing from a summer camp in Michigan, remembers: "For one fleeting moment, the moon landing was like the moon itself, raising a tide of joy, advancing a labor that birthed a sense of wonder." President Nixon called the flight "the greatest week in the history of the world since the Creation," and everywhere men and women congratulated themselves on the extraordinary feat. Command Module pilot Mike Collins has spoken movingly of what the landing meant to people around the world. When he and his crewmates went on an extended, multi-continent tour later that year, he was amazed by the jubilant crowds that turned out to greet them. And the most touching thing about the people who came to cheer, Collins suggested, was the way so many of them said, "We did it!"—acknowledging, as Armstrong had proclaimed, that it wasn't *Americans* who had walked on the moon, but human beings.

Bruce McCandless never lost his reverence for those days, and the socio-scientific supernova that Apollo represented. Even the flaws in the mission seemed to hold some minor message about human optimism and spirit, about moving so decisively that you're not quite sure where you are when you get there. It's not generally understood, for example, that no one on Earth knew exactly where the Eagle touched down. Collins, who remained in space, circling the moon as his comrades explored the lunar surface, tried to pinpoint the spot through use of a sextant but was unsuccessful. Another oddity: It was practically impossible to tell Armstrong from Aldrin on the moon, given that their suits were identical. Technicians wrote notes to themselves to add identifying stripes to space suits on future flights. It was all part of the atmosphere of euphoria and disbelief attendant on the landing.

My dad held all of his colleagues, not just his fellow astronauts, in high regard. But his special admiration was reserved for Neil Armstrong, that most elusive and phlegmatic of Apollo's heroes. Armstrong was some strange combination of magus, mute, and walking monument, a man who sometimes seemed to be observing the rest of the human race through the lens of a powerful telescope, baffled by our customs. He'd suffered through his share of hair-raising danger and miraculous escapes. He lost part of a wing of his F9F Panther during a Korean War combat mission and parachuted to safety. As a test pilot, he found himself miles off course when his X-15 aircraft, a plane built much like a missile, "bounced" off of Earth's atmosphere on its descent from a climb above 200,000 feet. While the aircraft did eventually fall low enough for Armstrong to regain steering control, the X-15 was out of fuel, and he found himself on a flight path into heavily populated areas of Southern California. With no viable options for landing, the young pilot was forced to make a long, agonizing glide back to Edwards Air Force Base, where he barely reached the landing strip. Gemini 8 almost killed him, until he figured out, in the middle of being spun into unconsciousness, how to stop the tumble that was turning his capsule into a blender. And in May 1968, just over a year before the launch of Apollo 11, he had to eject from the Lunar Landing Research Vehicle, the so-called flying bedstead, and parachute to a landing in a scrubby field just north of the space center. Bruised and bleeding from the mouth, he nevertheless dusted himself off, changed clothes, and headed back to the office to handle some paperwork.

People admired Neil for his even temperament and unassuming manner, even as some questioned how, exactly, a normal human being could function in the face of such hazards. My dad came to feel a greater bond with the man because Armstrong was not just a pilot or a scientist but something greater than both: an *engineer*. Armstrong could easily have ridden his fame for another twenty years after he landed on the moon, enjoying a sinecure in the astronaut office and taking whatever assignments he cared to while the public waited for him to appear and offer up gnomic wisdom regarding spaceflight, toothpaste, and whatever else he felt like talking about. But he got sick of the attention. Instead, he retreated to the University of Cincinnati to teach aerospace engineering. He shunned the limelight, declined to write a book, turned down any number of television appearances. To my dad, this combination of personal valor, intellectual

curiosity, and disdain for publicity made him the perfect space explorer—
the astronaut's astronaut.

APOLLO 11 IS A story that will one day assume the status of legend (and
possibly, if we're not careful, myth). In the spring of 2019, just a few months
before the fiftieth anniversary of the first moon landing, my wife, Pati, and I
saw the documentary movie *Apollo 11*. The film is magnificent, and I'm not
just saying that because Bruce McCandless shows up in Mission Control in
a couple of his crazy turtlenecks, looking like an extra from *The Mod Squad*.
We liked the movie so much that we decided to rent a theater and show it
to our friends. There were two teenage boys in the audience that day. The
mother of one of the boys apologized for asking to bring them. They like to
play video games, she whispered. Violent ones. But we wanted kids in the
audience, so we said that's fine, we're glad they're here.

Even after the movie started, those kids chuckled as they whispered
to each other, watching no-doubt-hugely inappropriate videos on their
phones and probably trying to figure out how to dispose of all the grownups in the world without getting grounded. But when they saw Armstrong,
Aldrin, and Collins suiting up on that sunny day in July 1969—the astronauts outwardly calm and contemplative but their eyes wide with a quiet
apprehension they couldn't quite mask—all talking stopped. There's a language in those expressions much deeper than words. The kids were silent
when the Saturn V rocket erupted in fire and thunder on the Cape Canaveral launchpad in front of them. And when Neil set the lunar module, a
spacecraft no sturdier than your average box kite, down on the surface of
the moon, with his fuel running out and the world holding its breath, those
two boys cheered like kindergarteners. It was fifty years later and we all had
tears in our eyes. It was beautiful. It was *electric*, and I thought, yes, we've
gotten it right. This is what my dad would have wanted to see.

7

TRY

STAYING HOME

Bernice McCandless spent most of her life as a dutiful spouse, overshadowed by her brilliant and occasionally famous husband, and a doting mother, way too fond of her impish daughter and doltish son. I loved her unreservedly, and I can say without exaggeration that we never in our adult lives had a fight. The only thing she ever asked of me was a grandchild or two. She never criticized or second-guessed me, and it took her death to make me realize what that really meant and how unconditionally her devotion was ladled out.

If you've ever read *The Giving Tree* by Shel Silverstein, you have a basic understanding of my mother. She was put upon, preyed upon, and drained of vitality on a daily basis by her family, who expected her to be all things to all people at all times. She wouldn't have had it any other way. She fed and clothed me and taught me how to play baseball in the backyard. When I was in elementary school, entranced by the contemporary soul stylings of The 5th Dimension, particularly their huge hit, "Workin' On a Groovy Thing," she sewed me a pair of red cotton pants with an elastic waistband and all the signs of the zodiac emblazoned on them. She rescued me from stick-wielding bullies—see zodiac pants, above—and forcefully argued the case to my grade-school teachers that I was probably intelligent despite obvious signs to the contrary. She embarked on a fifty-year publicity campaign to boost my ego so I wouldn't be crushed by my father's accomplishments like Wile E. Coyote under an ACME safe. When I was fourteen, she drove me on my neighborhood paper route whenever it was raining. And it rained *a lot* in those days. In July 1979, for example, Tropical Storm Claudette dumped more than forty inches of precipitation on areas south of Houston in only 24 hours. We still refer to that job as "Mom's paper route."

In the astronaut program, dads—and it was all dads back then—were frequently absent. Like military wives, astronaut wives were expected to keep the family together while their husbands did whatever it was they did on their frequent trips—trained at the Cape, met with local widget impresarios in Columbus or Fresno, studied volcanic rocks on geology field trips to Iceland and Kilauea. My mother and her peers were essentially asked to strangle their inner selves in support of their spouses, the program, and the red, white, and blue. Astonishingly, they did it. The astronauts climbed into their T-38s and shot off into the clouds. Their wives did the thankless, tedious work back in Houston. They cooked meat loaf and made tuna-fish casseroles and took kids to the Flight Clinic at the space center to see Nurse Dee. They helped with homework and taught their kids how to ride bikes. If that wasn't enough, they worried when their husbands flew and worried more when they attained that glittering and potentially lethal prize: a mission. I say worried *more* because once a man returned safely from space, there was another peril for his spouse to confront. The period immediately after a mission was, many astronauts wrote, a huge letdown from the anticipation and actuality of flight. The letdown led to depression, and depression to a desire for change. This meant divorce. Probably half of the early NASA wives were eventually discarded by their more glamorous spouses, and that prospect—the notion of being abandoned, left behind, with no money and no husband—preyed on my mother all her adult life.

"You think it's hard going to the moon?" one astronaut's wife famously asked a crowd of reporters gathered at her door. "Try staying home."

GROWING UP, IT SEEMED to me as if my father's family lived lives
of propriety and heroism, then wandered off to die with dignity, possibly
while wearing white gloves. I later found out this wasn't true, as of course
it couldn't have been. But it was my understanding at the time, and such
impressions are hard to shake. My mother's family, the Doyles, lived in
a different world, a place where work was a necessity but not a calling,
where friendship was more important than warships, where life came with
frequent opportunities for minor disaster. My dad's siblings are intelligent
and highly educated people. With the exception of his sister Rosemary,
though, who is a fixture at McCandless family gatherings, they seldom
socialize together. They prefer their independence. By contrast, my mother's family—only two of the kids are still alive—talk frequently. They swap
memories of their lives growing up as if the recollections were faded baseball cards. Sometimes the stats have rubbed off, and the images are all that
remain. The address of their childhood home in Roselle, 222 Lafayette
Street, is repeated like a mantra.

   Bruce and Bernice were an unlikely couple in many respects. You can see
it in snapshots of their upbringing. In a family portrait from the mid-fifties,
Captain and Mrs. Bruce McCandless sit composed and imperturbable in a
spotless living room, their four children arranged around them like museum
pieces, nattily attired and perfectly coiffed, gazing calmly into the camera
lens. In a Doyle snapshot, by contrast, the family has been momentarily cor-
ralled somewhere outside. My mother, age nine, grins as she tugs at her coat
sleeve. Her twelve-year-old brother, Charles, is tilted so far away from the
group, it looks like he's about to fall over. They're not posing; they're planning a mutiny. My grandfather Doyle stands largely occluded by his unruly
brood, a quizzical look on his face, as if he's not quite sure how he got there.

   You can also see the differences in my parents' personalities in their
handwriting. My mother's script was ornate and full of loops. It leaned

backward toward the past, as if it were having a hard time escaping its pull. Dad's was rushed, angular, occasionally illegible but tracking aggressively across the page. She was the left-handed, daydreaming baby of her family, beautiful but wistful and unsure of herself. He was the driven, right-handed, eldest child of a military clan, who kept his dreams on a workbench, where he could disassemble and spray them with WD-40 when they started to squeak.

It wasn't a perfect marriage. At various times it brought both my parents a fair amount of pain. But my mother was loyal and dutiful above all other things, and for her, commitment was nonnegotiable. She kept the marriage together despite long odds. She had to. Anything else might have killed her.

8

VOLUNTARY
COMPLACENCE

Boyhood for me was a perfect place. I had a gold Stingray bicycle and a brand new baseball glove. I could recite the Pledge of Allegiance from memory, and I was happy to say a prayer for the president. It's true that when the wind blew from the north we in El Lago could smell the hydrogen sulfide emissions of the petrochemical plants in Baytown, just a few miles away, an odor that suggested the devil was burning vinyl couches. But that meant nothing to an eight-year-old. The neighborhood kids—Matt and Curt, Mike, the two Dougs and I—constructed forts in the woods and threw bamboo spears at each other. We built rafts out of two-by-fours and cast-off pieces of plywood, tied five or six empty Clorox jugs to the keel, and set off to see the world, or possibly just Timber Cove, which was the ritzy subdivision across the lake where the glamorous astronauts lived.

If the wind blew hard enough, the water in the lake receded toward Clear Lake and Galveston Bay, and we could walk in the mud almost to the middle. We could, that is, if my mother didn't know what we were up to. If she did, she would race down to the shore and holler until we came back. She was convinced one of us would get trapped in the mud and drown, contract pellagra or cholera, or, even worse, get "worms"—ringworms, presumably—my mother's number-one worry about living in the South.

Our house was situated eleven feet above sea level on a lot near Taylor Lake. It was wet, marshy land, sediment so newly reclaimed from the sea that my first memory of the Manned Spacecraft Center is of the decorative rocks around the center's duck pond, the only stones for miles around. When Europeans first arrived, the area was dominated by the Karankawa, a tall Native American people with elaborate facial tattoos, expert archers who were alleged by the Spanish to eat the flesh of their enemies. The Karankawa were essentially exterminated by Texan and Mexican military actions in the nineteenth century. Galveston, called the Island of Doom by early explorers, eventually became a haven for pirates like Jean Lafitte and other outlaws; indeed, as late as the 1950s, the so-called Free State of Galveston was a noted gambling, bootlegging, and narcotics hub controlled by organized crime. Those days were long gone by the time we arrived on the Gulf Coast, but ghosts of that seedy glamour still whispered in the shadows on late summer evenings. Our little lake had reputedly been a hideout for Lafitte and his buccaneers in the early years of the American republic, and when I was growing up I would occasionally see treasure hunters prowling the shore, head down, following their metal detectors through the marsh grass.

My dad designed our house, and it bore his imprint. On the western side of the structure was a small, soundproof room that he used as his study, complete with wood paneling and plush scarlet carpet. He could close three separate doors ("hatches," he sometimes called them) for enhanced privacy. My bedroom had multiple electrical outlets, not yet the norm in those days, to allow me to conduct experiments in chemistry, physics, and electrical engineering—none of which I ever did, unless combining various distillates of alcohol counts as chemistry. There was a laundry room upstairs—more evidence of androcentric design—and a dressing room for my mom and dad downstairs that stood at the junction of the study, the master bedroom, and the bathroom. The bathroom was decidedly inelegant, so narrow that it had a naval feel.

Tracy and I slept upstairs. I can still remember my dad's bedtime routine: a tour of the doors downstairs, each of which he double locked, and then a sortie upstairs to turn off the lights, survey the guest room for prowlers, and say good night. He needn't have bothered checking on us. El Lago was a relentlessly safe place. One of the bigger controversies to hit the neighborhood was when the city decreed that no cars were to be parked on the street overnight. A notice went out through the local constabulary, which posted it on the windshields of errant automobiles. The flier announced the no-street-parking policy and added that the police department appreciated our voluntary complacence. We figured they meant *compliance*, but it didn't much matter. In El Lago, they got both.

THERE'S A SCENE IN *First Man*, the Neil Armstrong biopic, in which Armstrong's wife, Janet, persuades him to sit down with his two sons for a serious discussion. He needs to be straight with them, she says. He needs to explain that he's going into space and may not be coming back. Neil reluctantly complies. He tells the boys he's going to be gone for a long time and waits for the gravity of the situation to sink in. His younger son looks up, serious and thoughtful. "So you won't be home," he says, "for my swim meet?" It's a scene that gets a lot of things right. Few kids are capable of understanding what spaceflight means. I know I wasn't. I watched news coverage of Apollo 11 at our community pool. It was cool. So was the high dive, which I'd finally overcome my fear of that summer.

Still, there were elements of that life I enjoyed, as long as no one was watching. The Manned Spacecraft Center was mostly an open field at that time, speckled with pink Texas primroses in the spring and waves of swaying spear grass when the weather got hot. Its low-set white buildings stood together well to the north of NASA Road One, anonymous and unremarkable, as if someone had set a herd of parking garages loose on the coastal prairie to graze. A trip to the space center for a visit to the astronauts' gym—with its exotic, seemingly bottomless supply of giant, citrus-flavored salt tablets and its neatly folded stacks of blue shorts and shirts—meant entering through the east gate. Here the uniformed security guard would give my dad a casual but significant salute, two fingers to the forehead, and Dad would respond in kind. Never mind that he wasn't in uniform.

Evidently Bruce McCandless was someone you saluted anyway. The two gestures, understated and unacknowledged, carried whole pages of information: *I know you; you know me; we both understand the situation around here and everything is copacetic at the moment, but we'll be watching for Soviet spies and the like while you're inside the center, attending to some pretty damned significant space-type business.* This casual system of physical semaphore—no unnecessary effort or information—seemed to me the embodiment of how to be a man. Saying my dad was an astronaut was a tough hand to beat. Your father could be a cop, an engineer, even a pilot, but nothing was cooler than an astronaut. True, there were years when I couldn't answer the inevitable follow-up question, "Has he been up in space?" the way I wanted to (fifty times, and he wrestled an alien), but even that shortcoming was kind of cool. I mean, I had an astronaut for a dad, and he hadn't even bothered to go into *space*.

Because so many of El Lago's breadwinners were involved in the space program, drawing either government or government contractor wages, our little suburb had an egalitarian feel. There were nicer houses—*doctors' houses*, people said—that lined the shore of Taylor Lake, but there weren't any mansions. No one had a Porsche or a Ferrari parked in the driveway. Only one of my friends had a pool. No matter who you were or thought you were, kids were valued primarily as a source of cheap labor. We washed cars, swept driveways, and, in my case at least, mowed the lawn.

We had a big, lush, man-eating lawn. To my eyes, it was about the size of New Hampshire. And size wasn't even the main problem. We lived on the Gulf Coast, after all. Every April the sun would descend from the heavens like an angry god to float just above the refinery stacks in Baytown and superheat the atmosphere to a consistent ninety-seven degrees, just warm enough to entice the rich aromas of benzene and toluene from the steel tanks. This, combined with the prevailing 80 percent humidity, meant that the St. Augustine grass grew at alarming and inexhaustible rates. As did the fire-ant beds. And the swarms of ravenous mosquitoes, no bigger than nail clippings but formidable when swarming. There was a section of our side yard that was slightly lower than the rest and frequently wet as a result. This is where the crawdad chimneys were located—and the crawdads, presumably, though they always had enough sense to stay underground while I was mowing. A crawdad is a reddish brown, lobster-like creature that thrives in the warm, rainy climate of the Gulf Coast. In dry weather it burrows into the earth to find moisture, and the soil it excavates forms a gummy funnel sticking up to mark its abode. The burrows weren't a serious menace, but they were just glutinous enough to clog the mower and slow me down. Thus there would be days when, by the time I finished the front yard, both side yards, and the backyard, it was almost time to mow the front yard again. I was like Sisyphus, only instead of a huge rock I was pushing a seven-horsepower Craftsman lawnmower and being chased by wasps.

Over the years I attempted to donate approximately 400 square feet of property to our neighbors in an attempt to get someone else to mow the damn grass. All my donations were politely declined. I feigned illness: exotic, domestic, and mental. I removed screws and bolts from the mower and promptly lost them. But here's the problem with having an engineer for a father. Equipment failure was not an option. Bruce McCandless could repair anything. So even if the lawnmower wouldn't start, there was no cause for celebration. Maybe I wasn't pulling the cord hard enough. Maybe the fuel line was clogged—easy enough to remedy. Or maybe someone had removed the spark plug and hidden it under the left rear tire of the Suburban until the Astros game was over and the sun was no longer an imminent threat to my prepubescent health and sanity. No problem there either—my dad always had a few extra plugs lying around, and spark plugs were an easy fix.

"Onward," he'd say, grinning, as the mower roared to life.

"Sideward," I'd grumble as my afternoon plummeted earthward.

DESPITE THE EXCITEMENT OF the Apollo program, these years were bittersweet for my dad. His grandfather Byron died in 1967. His *father*, the first Bruce, died in 1968 after losing his long battle with the progressive paralysis that had started sixteen years earlier. My grandmother Sue spent a long time dealing with her husband's illness. Not long after he passed, she married an Atlanta businessman, a longtime friend, and tried to restart her life. There must have been a world of unspoken sentiment in our house about the death and quick remarriage, but it was kept well hidden, and I had no interest in those days in life behind the curtain. My father doubled down on his work with the jetpack, simulating the experience of maneuvering free and untethered in space for hours at a time. I sometimes wonder if these hours were a refuge for him, a way to escape reminders of his dad's grim imprisonment in his own body. Sue McCandless died in 1969, followed shortly thereafter by *her* mother, my great-grandmother Sue Bradley. Factoring in my grandfather and great-grandfather McCandless, this meant my dad lost four of the people he loved most in the span of four years. But he never spoke of it to me—even to the *adult* me. As far as I can recall, none of these events was ever mentioned in our house.

Looking back, this seems like the first manifestation of a trait that would become more pronounced as the years went on. My father disliked discussing certain subjects, and, as smart as he was, he was not immune to superstition. For example, if there was a hurricane in the Gulf of Mexico threatening to make landfall in Texas, we were forbidden to say its name. Naming it, I suppose, might have summoned it. The subject he disliked most was death. Maybe you have to ignore the idea of your own mortality to be a fighter pilot. Whatever the reason, Bruce McCandless steadfastly avoided all explorations of the topic. It was the Great Mystery, the Red-headed Gunman, unwelcome because it was beyond his control. There is a simple rule in whitewater kayaking: Look where you want to go. Your body and your boat will naturally move in the direction your eyes are pointed. Looking at the rock in the middle of the stream makes it more likely you're going to hit it, which is exactly what you don't want to happen. My dad applied a similar logic to death. He refused to focus on it.

Once when I was at home during college, I read my parents one of Donald Hall's poems. In it, the narrator talks about holding his infant son in his arms and realizing that he himself is starting to die. It's a short poem, brilliant and sharp, a punch in the gut. When I finished, I looked up for a reaction. My mother gazed off into the middle distance, sobered by this flash of forever.

"A little morbid," said my dad, as he wandered off to make himself a snack.

BRUCE McCANDLESS'S PRIMARY BEHAVIORAL characteristic was an obsessive need to be productive. His go-to phrase, suitable for almost all occasions, was "Onward." (In times of extreme duress, he might opt for the more hortatory "Hang in there" instead.) When he wasn't out of town, he was likely poring over technical papers in his study or banging, drilling, or rewiring something in the garage. He hated inactivity. There was always something he needed to fix or figure out, read or revisit. He disliked family reunions because they featured lots of what he called "milling around"—*i.e.,* unnecessary talking. Though he was more social than my mom, especially when his career prospects were involved, he still preferred to "make an appearance" at splashdown parties and award ceremonies rather than stay for extended periods.

He built our first color TV, a giant Heathkit model that we used till the knobs fell off and the picture was mostly a murky green, which actually lent an air of additional creepiness to *The Outer Limits*. He helped me construct an electric motor for my second-grade science project, a contraption that won a blue ribbon and saddled me with a long-lasting hangover of embarrassment because I knew I never could or would have built it myself. He cut and stained and installed the bookshelves in our family room. He assembled a shortwave radio receiver for me—another Heathkit—and fashioned an elaborate Reddy Kilowatt costume with a ten-watt nose bulb for Halloween one year. But his excellence could be an imposition. Because my mother was a doting parent, I felt like the king of the house when my dad was gone, and I wasn't always thrilled to see him return. When Dad was traveling, Tracy and I ate TV dinners and watched *The Dick Van Dyke Show* and *Adam-12* in the family room. When he came home, we had to sit up straight at the dinner table, eat our vegetables, and generally relearn that we were second-class citizens.

But there were consolations to his return as well. My mother relaxed when Dad was around. There was a sense that the captain was back at the helm. A treat back then was an after-dinner trip to the Baskin-Robbins on NASA Road One. My father would have nothing to do with Black Walnut or Here Comes the Fudge. He much preferred Mint Chocolate Chip, with its science-fiction green hues and vaguely medicinal taste. Once the dessert melted enough to liquefy, which in Houston took about twenty seconds, he ate the ice cream by biting a hole off the end of the cone and sucking it out through the bottom. It was his engineer's solution to sticky fingers and an unstable dessert bolus. It wasn't a good look, generally speaking, but it worked.

I wish I could have absorbed a portion of that engineering intellect— maybe not the ice cream-sucking part, but something. But I veered from the course. I lost my enthusiasm for science, and Mrs. McGregor's second-grade forced march through advanced subtraction left me cowering in the weeds of mathematical incompetence. Even at a young age I frequently got in trouble for talking back. It got worse as I got older. But we're not there yet. Leave me here in 1969 for a while. My Sony transistor radio is tuned to KILT 610 AM, and Shocking Blue's "I'm Your Venus" is murmuring from the tiny speaker. Venus, who looks a lot like Raquel Welch, is as much idolatry as I'm prepared to allow in my bedroom. I am the neighborhood Strat-O-Matic Baseball king. Sherman tanks guard my bookshelves. I have a beautiful mother who loves me and a dashing, good-looking dad who's currently in Iceland or Argentina or somewhere similarly outlandish. In my mind it's not just my parents who are looking out for me. It's NASA in general, and NASA, as everyone knows, is America itself, powerful and infallible. In my neighborhood, kids and dogs are everywhere, and summer evenings are as warm as my mother's smile. Life is good, I think. God is watching. And if He's not, the satellites are.

9

APPALACHIA
UP ABOVE

I t was bound to happen. It was human nature, after all, caught up in the iron laws of anticipation and release. The enthusiasm of the space race ebbed as several successful but less momentous moon landings followed Apollo 11. The hangover came quickly. Something about men hitting golf balls on the moon and gadding about in the dune-buggy-like lunar rover seemed to trivialize the enterprise. Leave it to Americans, some said, to bring a *car* to the Descartes Highlands.

Even the astronauts knew the glory days were waning. One wag drafted up an Apollo 13 flight plan in early 1970 that called for the crew to "use a special hydraulic shovel to dig a foxhole, then fill it up again, to demonstrate man's capacity to develop the moon. After reading excerpts from the Book of Deuteronomy, Eisenhower's memoirs, and the Wilmot Proviso, [the astronauts will] speak personally with many governors." NASA was a victim of its own success. It had made the impossible routine. MAN'S LEAP FOR MOON ISN'T LIKE IT ONCE WAS, wrote the *Miami Herald*, commenting on the decline of the boozy excitement that had once attended Apollo launches. The most tangible acquisition of the flights was several boxes of moon rocks, which lent credence to the notion that perhaps the results of the program weren't as great as the effort that went into it. Congressional attention wandered, budgets declined, and eventually the planned Apollo missions 18 through 20 were canceled, leaving (incredibly!)

three ready-to-fly Saturn V rockets stranded on earth. NASA wasn't sure what to do with these marvels of engineering. It was as if the ancient Egyptians had built too many pyramids, and started wondering if maybe the Mesopotamians could use one.

The last lunar flight, Apollo 17, was scheduled to take off from Cape Canaveral on December 7, 1972. It would be commanded by "spacewalk from hell" veteran Gene Cernan, with Ron Evans and geologist Jack Schmitt as crew. Not only was it scheduled to be the final Apollo launch. It was also going to be the only Apollo *night* launch, and my father was determined to get us to Florida to see it. So Dad drove us the thousand miles from Houston to Cape Canaveral in his Volvo and a very bad mood, east through the industrial killing fields of Beaumont and Port Arthur and then into the nighttime swamps of southern Louisiana on I-10, bird calls filling the cabin of our brand-new four-cylinder Swedish economy vehicle.

I woke the next morning as we passed highly suspicious archetypal alligator farms in the Florida Panhandle, lurid, filthy, and dangerous, which I desperately wanted to see but couldn't because there wasn't a Stuckey's for another 64 miles and Dad wasn't interested in stopping unless someone died or the car blew up, and possibly not even then. Then moving inland we veered south into the gut of the pine-treed peninsula, tracking for the Atlantic Ocean. I spent the latter half of that long drive bored, still peevish about not getting to see some genuine reptile wrestling, as promised by the thirty-odd highway signs we'd passed, and unprepared for what happened once we made it to the Cape. *Darkness.* Hours of it. And waiting. Adults stumbling around like they usually did, saying pointless things for no particular reason. Just as I was about to set out walking the long miles back to my beloved alligator farm, where I hoped to sign on as an apprentice, my mom came running to grab me. There was a quickening in the crowd. It felt like being at a wedding just before the bride appears. The voices around me grew fainter but more urgent, and my dad said prayers to his camera. Mom held Tracy's hand. And then it happened. The flash. The ground to the east lit up as the big rocket ignited. We were three miles away, so we couldn't hear anything yet, but the rocket started climbing, inching up from the pad, and then the noise washed over us, a low-pitched rumble so huge and intense I felt unable to move. The air crackled like a whole sheet of Black Cat firecrackers. The earth was on fire as bright as July and the ground trembled with the thrust of von Braun's massive white sky spear, all seven million pounds of it, as it ripped off the garments of gravity. We couldn't help it. We cheered. One observer pointed out that it would have taken 30,000 strong men to raise that rocket an inch off the ground. Now it was rising like a self-made sun, levitating like a minor god, a vision of technological rapture headed for an object in the heavens 240,000 miles away. The rocket dwindled in the night sky. It turned into the flame of acetylene torch. It was a fiery shuttlecock. It was the point of light on a dying TV. And then it was gone.

It took a long time for the crowd to disperse. No one wanted to leave. We gazed at each other with newfound respect, proud that we, or people who looked and thought a lot like us, who spoke the same language we did, had just done something almost incomprehensible. Forget the alligator farm. What could possibly top this display of American precision and power? Apollo was over. So was the excitement. What happened next was a slow-moving, script-less disaster: the rest of the decade.

WHEN THE APOLLO PROGRAM ended, so too did Bruce McCandless's chances of walking on the moon. It was a crushing disappointment. Not only had he missed out on the actual Apollo missions; in all the talk about how the canceled missions might have been crewed, his name rarely came up. In other words, he wasn't even given a slot on an *imaginary* flight.

It wasn't a fallow period for space exploration generally. Indeed, the seventies saw the launch of a number of ambitious unmanned probes and satellites. The Soviets landed a rover on the moon and put probes on Venus, though the little machines didn't last long in the planet's extreme heat. The United States placed Mariner 9 in orbit around Mars in 1971, sent two

Viking landers to Mars in 1975, and launched its two Voyager probes in 1977. Voyager 2 took photographs and measurements of the solar system's planetary ice giants, Neptune and Uranus, and discovered six previously unknown moons and a planetary magnetic field around Neptune. The Voyagers are, in fact, still traveling. As a NASA website points out, as of this writing, Voyager 1 is three billion miles outside the heliosphere and transmitting data back to us by means of a signal that takes twenty-one hours to reach Earth—about the same length of time Armstrong and Aldrin spent on the moon. Voyager 2 is almost a billion miles into interstellar space. Each probe carries a golden phonograph record that contains images and sounds, including greetings in a number of languages, meant to provide an introduction to Earth and its residents. For those advanced extraterrestrial civilizations that have moved on to eight-track tapes, graphic instructions illustrate the correct way to access the information contained on the disk. And unlike old Rolling Stones albums, each record cover contains a small amount of uranium-238 with a radioactivity of about 0.00026 microcuries, which could be used by aliens to determine the amount of time the disk— and thus the satellite that carries it—has been traveling.

There were still opportunities for manned spaceflight in the seventies as well, though they weren't as exciting as the moon missions. The Apollo Applications program, renamed Skylab in 1970, was up next on NASA's agenda. Though the program was makeshift from the start, criticized by some as simply a way for NASA to keep money coming in the door as the agency worked to develop its real ambitions—a space shuttle and a permanent space station—Skylab involved real science, real engineering, and, in the end, real fortitude on the part of its inhabitants.

If you're unfamiliar with the story of America's first space station, you're not alone. The project has always struggled for respect. But Skylab, America's VW microbus in the sky, is definitely worth a few minutes of your time. I know, because it took a few years of my father's. The most important reason for this is that in January 1972 NASA named the prime and backup crews for the first manned Skylab flight, designated as SL-2. The prime crew consisted of Apollo veteran Pete Conrad as commander, along with pilot Paul Weitz and Navy physician Joe Kerwin. Bruce McCandless was named to the backup crew, along with Rusty Schweickart and Story Musgrave.

As the backup for Weitz, my father was just an erratic heartbeat away from space. Photographs from this period show him in high spirits. He trained with the prime crew and basically did everything they did. He sampled new and supposedly more palatable Skylab foods, participated in the mission's crew compartment fit-and-function test at McDonnell Douglas facilities in St. Louis in June 1972, and completed altitude chamber tests of the Skylab 2 command module with Musgrave and Schweickart in January 1973. He practiced splashdown procedures and retrievals from the command module by Navy frogmen in the Gulf of Mexico. He was elated to be so close to going up. As it turned out, though, prime crew members Conrad, Kerwin, and Weitz remained perfectly healthy, which meant Dad and his backup crewmates remained earthbound.

The launch of the first Skylab mission, SL-1, took place on May 14, 1973, just a few months after the final flight to the moon. There were problems from the start. A Saturn V rocket boosted the actual Skylab—the cylindrical lab itself, unpopulated at this stage, and fashioned from the Saturn V's third stage—into Earth orbit. Sporting four fan-blade-like solar arrays meant to service the station's Apollo telescope mount, Skylab resembled a giant windmill in space. It was also designed to have two large rectangular solar arrays, one on each side, as if it were gliding on a pair of stubby wings. Unfortunately, the station lost its micrometeorite/solar-protection shield and one of its two big solar array "wings" as a result of several mishaps shortly after takeoff. The second solar array appeared to be intact but failed to deploy—that is, spread out—as designed.

These were crippling blows. Skylab was designed to operate largely on the power generated by the two big solar arrays. The micrometeorite shield was meant to provide protection against random space junk crashing into the station, yes, but it was also designed to shield Skylab from solar radiation, keeping the station cool enough for human beings to live and work inside. Now the shield and one of the arrays was gone, with the remaining array nonfunctional. As a result, temperatures aboard the spacecraft climbed to 130 degrees Fahrenheit. The station was uninhabitable under such conditions. In short, Skylab seemed likely to be a total loss—an embarrassing and expensive mistake.

SL-2, the first manned, or, in NASA's modern parlance, *crewed*, mission, was originally scheduled to take off the day after Skylab 1. Once NASA became aware of the damage to its new space station, however, the second launch was delayed for ten days. Engineers and technicians at JSC and Marshall Space Center scrambled to come up with ideas for fixing the crippled satellite. As ideas were developed, astronauts took to Marshall Space Center's Neutral Buoyancy Simulator to try out and rehearse the procedures that would be needed to remedy the problems. Bruce McCandless was assigned to work on ways to free the balky solar array, which seemed to be operable but which was being held closed by what appeared to be a thin strip of metal. He worked with the A. B. Chance Company of Centralia, Missouri, to procure a pair of utility lineman's tools—a cable cutter and a "universal tool with prongs for prying and pulling"—both of which could be mounted on a three-meter pole. These tools and the pole were added to the equipment manifest and carried into space when Skylab 2 launched on May 25, 1973. Pete Conrad and Joe Kerwin used the tools while standing on the satellite as it cruised through the heavens, 270 miles above Earth, at 17,000-plus miles per hour. They cut a piece of aluminum that was keeping the one undamaged solar array from deploying. Even after the aluminum strap was cut away, though, the recalcitrant solar array remained stuck, with one of its hinges essentially frozen in place. In a tense and wholly improvised procedure, Conrad and Kerwin had to use a rope to yank the array loose, after which effort the astronauts briefly soared off into space before being restrained by their tethers.

Skylab 2's crew also deployed a sort of Mylar "parasol" to shield the orbital workshop from solar radiation, a temporary fix that was improved a few weeks later when Skylab 3 astronauts Jack Lousma and Owen Garriott installed a sturdier twin-pole solar shield instead. Even with these repairs somewhat miraculously effected and the station restored to more or less full functionality, the spacecraft was lopsided, like a mobile home with a broken window and an inelegant tarp that looked like a giant bandage. There was something hillbilly about the whole affair, a perception confirmed when Skylab unceremoniously fell out of the heavens in 1979 and scattered itself across Western Australia.

Space enthusiasts justifiably cite Apollo 13 as the most spectacular example of a space repair. There were three lives at stake, after all, and the world was watching. But the resurrection of the ailing Skylab by the crew of SL-2 surely comes a close second. Conrad and Kerwin's heroics notwithstanding, though, the American public was largely unimpressed by the flights. It wasn't just that the country was still recovering from the techno-rave that was Apollo 11. Skylab simply wasn't pretty. The dazzling high-definition images captured by astronauts working outside the space shuttle were almost a decade in the future. To make it worse, the aims of this next phase of space exploration—extended stays in low Earth orbit and solar observation—seemed fussy and diffuse by comparison with walking on the moon. Finally, there was other news occupying the nation's attention in the early seventies, and very little of it was good. There was the single biggest political scandal in United States history, for example.

In March 1973 a former CIA operative named James McCord wrote to Federal Judge John Sirica, claiming that the defendants in the Watergate Hotel burglary case, himself included, had perjured themselves in response to pressure from highly placed individuals in the United States government, and that he, McCord, faced retaliation if he disclosed all that he knew about the origins and intent of the burglary. McCord subsequently cooperated with prosecutors. The Watergate hearings that year were viewed with appalled fascination by millions of Americans. Former White House aide Alexander Butterfield famously disclosed that summer that President Nixon had installed a system for tape-recording all conversations in the Oval Office. Special prosecutor Leon Jaworski eventually obtained some of these tapes, which revealed a venomous and bigoted behind-the-scenes Nixon that few Americans cared to support. The president would resign from office in August 1974.

News also broke in 1973 that Nixon's vice president, Spiro Agnew, was facing a federal bribery investigation. Though he strenuously denied the allegations against him, Agnew eventually resigned from office, pleading guilty to one charge of tax evasion rather than facing certain impeachment. And the hits kept coming. The dollar was devalued. Native American activists occupied Wounded Knee, and American troops continued to return home from Southeast Asia, signaling the last stages of our bloody and disillusioning involvement in Vietnam. They were strange times in America, days of recrimination, self-doubt, and guilt. Skylab, it turned out, wasn't exactly the prescription for relief.

SCIENTISTS LIKE WERNHER VON BRAUN had for years advocated placing permanent space stations in Earth orbit. He thought it was important to keep NASA working, even if the goals of that work were somewhat unrealistic at the time—and, later, became *hugely* unrealistic as a result of cuts to NASA's budget. Yet aside from a way to keep Apollo technology and manpower online, Skylab was also an obvious response to the Soviets' manned space station, Salyut 1, which had been deployed in 1971. Three cosmonauts had visited the station in June of that year and stayed for twenty-eight days, setting a record for the length of their stay in space. The crew didn't have long to enjoy their achievement, though. During their return to Earth, their Soyuz spacecraft malfunctioned, and the resulting depressurization of the vessel killed all three men.

In the competitive sense, Skylab was a quick success. SKYLAB CREW TAKES OVER SPACE ENDURANCE RECORD, crowed one American daily on June 19, 1973, after the first Skylab crew's stay in space surpassed the Soviet mark. But that was just the start. Altogether, three crews of three astronauts visited the American space station in 1973 and 1974, spending a then-staggering total of 171 days in orbit. The totals seem paltry now, but only because Skylab astronauts paved the way for their successors' achievements. More recently, Scott Kelly spent 340 days in orbit on the International Space Station. Christina Koch set a woman's record for 328 days in 2020. And Russian cosmonaut Valery Polyakov spent a whopping 438 days in Russia's *Mir* space station in 1994 and '95.

The nine astronauts who visited Skylab—and the first crew in particular—spent a significant amount of time repairing it. All the while, the astronauts were themselves the subject of the station's central experiment, a fine-grained survey of changes in human physiology in space. NASA went to considerable lengths to try to make the spacecraft comfortable for the extended stays the crews had signed up for. There was an exercise bike to keep hearts and muscles strong, a not-very-efficient hot-water shower, and a wardroom for communal dining, along with three private chambers for sleeping, reading, and general relaxation. Some of the amenities worked fine. Others didn't. The shower was a flop, for example, but the viewing portal in the wardroom turned out to be a big hit. It was all part of the experiment. Some of the lessons NASA learned during the Skylab missions, like the importance of exercise during space travel, are now incorporated as everyday features of life on the International Space Station.

Even after his chance for a space station flight passed, Skylab held special significance for my father. All through the excitement and eventual decrescendo of the Apollo program, Bruce McCandless had continued to work on improving the Astronaut Maneuvering Unit. He'd eventually realized that Gene Cernan's difficulties in getting to and into the AMU on Gemini 9 had soured NASA on the device. Nevertheless, he became increasingly convinced that the unit could serve a crucial function in supporting satellite deployment and repairs, space-station maintenance, and possibly even crew rescues on future missions. One of the toughest parts of getting his project to fruition, he sometimes said, was convincing NASA that the jetpack idea still had merit. He and a man named Ed Whitsett embarked on a secular crusade of sorts, touting their engineering improvements and testing data to whomever would listen. Dad sometimes quoted Wernher von Braun, who wrote: "We can lick gravity, but sometimes the paperwork is overwhelming." Slowly, very slowly, the agency started listening.

Bruce McCandless kept in his files a letter from Rusty Schweickart dated June 19, 1970. The missive demonstrates that he and Whitsett were making progress in their jetpack promotional campaign. The newest version of the machine, a successor to Gene Cernan's AMU called the ASMU, for Automatically Stabilized Maneuvering Unit, had been officially cleared for testing on board Skylab. The experiment was designated M509, and the iconoclastic, intellectually restless Schweickart was getting excited about the prospect of untethered flight. "For the advantages in evaluating the precision maneuvering capability of M509," Schweickart concluded, "I feel we ought to plan some free flight-suited [that is, pressure-suited] runs. There appears to be no significant reason not to do so."

As it turned out, Skylab provided the first opportunity to test in microgravity, side by side, three different and competing methods of proposed extravehicular propulsion. It was a sort of Jet Power Pageant, and featured:

1. foot-based propulsion, the so-called "rocket boots," designated as Experiment T020, in which thruster nozzles were located beneath, and controlled by, the astronaut's feet;

2. a version of the Hand-Held Maneuvering Unit (HHMU), a sort of two-pronged metal wand with small thrusters, used by astronauts Ed White and Michael Collins during the Gemini program; and

3. the new and improved version of the Air Force-designed AMU jetpack, now called the ASMU. This was the machine Bruce McCandless, Ed Whitsett, and a NASA engineer named Dave Schultz worked on through the late sixties and early seventies. It had thrusters built into the jetpack and could be operated in three different control modes.

Each of the three devices, which were fueled by compressed nitrogen gas contained in a bulbous metal tank, or "bottle," was subjected to hours of test "flights" inside Skylab's cavernous orbital workshop during the second and third crewed missions. The rocket boots, which utilized eight gas-jet thrusters, were thought to have one big advantage over the other methods:

namely, they would leave the operator's hands completely free for performing extra-vehicular tasks. While propulsive footwear worked fine for Iron Man, Skylab testers gave them low marks because they had limited lateral thrust ability; their thrusters didn't fire in pairs to dampen unwanted "translations," or movements; and the subtle movements necessary for piloting the device by foot rather than hand seemed unnatural and unlike the ways the astronauts were used to controlling the machines they flew. It was, said Al Bean, "different from flying a spacecraft, or an airplane, or any other type of vehicle that has evolved from 70 years of flight."

The HHMU was basically a point-and-shoot device requiring only one hand to operate. Though simple to use, it was also imprecise. It had only two thrusters and required more or less constant attention to ensure that activation of one thruster or the other was aligned with the astronaut's center of gravity, to avoid unwanted changes in translation or attitude. Bean said: "Not only did I have difficulty putting it where I wanted to, I didn't know exactly where to put it. My [center of gravity] was not where I thought it was. But most of all, I didn't know what the suit effects were going to be, and nobody else is going to know either until their first EVA. Then they're going to find out what they thought they could fly has turned into a bucket of worms."

The ASMU jetpack, M509, had three modes of operation. The first, direct or "manual" control, basically meant that the operator was responsible for all aspects of where he went and how he was aligned with his target when he got there. In other words, the astronaut had to manually fire his or her nitrogen jets in order to travel anywhere or to stay upright or oriented as desired. The second mode was called "rate-moment gyro" and used integrated strap-down gyroscopes to control and stabilize relatively large deviations in attitude. The rate-gyro system, according to *Aviation Week & Space Technology*:

> Permitted the astronauts to digress from their initial attitude with a 4-degree deadband, with a 2-degree/second deadband established for rates. Use of such a system permits the astronaut to move around somewhat in the unit without the system automatically firing thrusters

to hold a steady position. If the deadbands were exceeded under the rate gyro mode, thrusters would fire automatically to bring the unit back to tolerable limits.

The third and final mode of ASMU operation was known as the "control moment gyro," or CMG, mode. It used the unit's three pairs of gyroscopes to counter disturbance torques and basically held the astronaut in a preset attitude by use of *automatically* firing thrusters to counter any movements deviating from such attitude.

The ASMU was flown in various modes by all three astronauts—Al Bean, Jack Lousma, and Owen Garriott—on SL-3 in 1973, and by Gerald Carr and Bill Pogue on SL-4's 1973-74 mission. All five gave the machine high marks. Bean said that "from a pilot standpoint, M509 was extremely comfortable," and "I would be willing to take [it] outside untethered the way it is now." Significantly, he also said that the unit would have been useful for inspecting damage to the Skylab after the SL-1 launch. Scientist-astronaut Owen Garriott flew the ASMU despite having no advance preparation for the task. He was surprised to find that no training was necessary. In fact, he liked piloting the ASMU so much that he asked if he could have some additional time.

It was a watershed moment for the jetpack, which emerged as the clear winner of Skylab's Jet Power Pageant. In a report signed by Bruce McCand-less, principal investigator Ed Whitsett, T. B. Murtagh, and J.T. Josephson, the authors summarized the favorable results and concluded that "an operational astronaut maneuvering unit can and should be developed for the Space Shuttle era to support a variety of EVA tasks such as shuttle support (inspection of the exterior, various contingencies such as stuck-open doors, manipulator assist), payload support (deployment, module replacement, maintenance/repair, retrieval), and crew transfer for a rescue situation ... With the transition from a research test unit to an operational EVA maneuvering unit, man's capabilities in orbital operations will take on a new dimension."

NASA agreed. As a result of these successful—if fully confined—M509 flights, the agency committed to development of a jetpack that would aid astronauts in the space-shuttle era to exit the mother ship and move around freely and untethered in space. With the conclusion of the Skylab program in early 1974, a second door into space slammed shut in Bruce McCandless's face. But the success of M509 suggested that maybe, just maybe, another one was opening.

MUCH OF THE REST of the astronauts' time on the Skylab missions was taken up with a wide range of experiments—the nation's high school students competed to get their proposals on board, and the best of these were incorporated into Skylab's task list—and with monitoring and recording solar activity. Astronauts captured important information about solar flares and other phenomena produced by our local star; indeed, data gathered by Skylab was the basis for work that won one scientist, Richard Giaconni, a Nobel Prize in Physics in 2002 for studies of X-ray astronomy. Along with study of the sun and physiological monitoring of the astronauts, Skylab had a third major focus: Earth. The astronauts were able to capture a stunning number and variety of images of our home planet through sustained, high-quality photography. As a NASA publication puts it:

> The more than 40,000 photographs made of the Earth and the thousands of observations recorded on miles of magnetic tape provided a mass of data . . . of great value to those involved in improvements of agriculture and forestry, geological applications, studies of the oceans, coastal zones, shoals and bays, and continental water resources, investigations of atmospheric phenomena, regional planning and development, [and] mapping and further development of remote sensing techniques.

Taken under the rubric of the Earth Observation Program, these images have ever since provided an important baseline for examining environmental and geophysical changes on Earth. Due perhaps to the nation's post-Vietnam fatigue, the astronauts were forbidden from calling the subjects of their photographs "targets," as the word sounded too militaristic. And because it was an age of détente in the long war between the West and the East, they were also prohibited from taking photographs over the USSR and China, though they weren't, of course, restricted from looking. SL-3 pilot Jack Lousma remembers gazing with particular interest down toward the Soviets' highly restricted Baikonur launch complex.

In gathering these images, the Skylab astronauts laid the groundwork, so to speak, for one of the most important of NASA's scientific endeavors—the continuing study, not of space, but of *Earth*. Many writers credit Apollo 8 with jump-starting the American environmental movement by capturing images of our little globe in a single frame, dwarfed by the immensity of the cosmos. The clear message was that our home planet is a lonely miracle, deserving of attention and care. Skylab came online as the environmental movement was reaching full strength. In the 1972 film *Silent Running*, Bruce Dern's character, possibly driven insane by the disembodied voice of Joan Baez following him through the corridors of a gigantic space warehouse, sacrifices himself to save a fragile forest from the callousness of corporate bureaucracy. The movie *Soylent Green* hit theaters a year later, right around the same time Skylab was launched, and forcefully warned Americans about the hazards of pollution, climate change, and deceptive packaging. Books like *The Population Bomb*, *Small is Beautiful*, and *Desert Solitaire* struck similar notes of admiration for the natural world and anxiety about the ways in which humanity and technology were altering it. In December 1973, as the third crewed Skylab mission orbited overhead, Congress passed the Endangered Species Act, arguably the most important piece of environmental legislation in the nation's history. It therefore seems appropriate that the three Skylab crews spent considerable time photographing, and indeed just *gazing at*, the blue jewel passing beneath them.

The agency's interest in earth science has never really subsided, though sometimes it seems as if the public's has. While the scientific freak flag that was Skylab is no longer flying, NASA continues to provide some of the best and most complete atmospheric and oceanic data about our planet. And the agency doesn't mince words. On its Global Climate Change: Vital Signs of the Planet web page, NASA says: "The heat-trapping nature of carbon dioxide and other gases was demonstrated in the mid-19th century. Their ability to affect the transfer of infrared energy through the atmosphere is the scientific basis of many instruments flown by NASA. There is no question *that increased levels of greenhouse gases must cause the Earth to warm in response.*" Such information can be seen as a direct legacy of Skylab. Less obvious but at least arguable is the proposition that Skylab restored America's admiration for science after years in which science's work, as embodied by "advances" like DDT, Agent Orange, napalm, and nuclear weaponry, was seen primarily as a tool chest for military and industry interests working toward the long-term detriment of the human race and its home planet.

AS JOE ALLEN COMMENTS in his book, *Entering Space: An Astronaut's Odyssey*, the men who went up in Skylab complained about the station's "poorly designed experiments, faulty equipment, awkward toilet facilities, poor lighting, drab clothing and bland food." The space travelers on the Mercury and Gemini missions hurtled through the cosmos with bright eyes and chins as smooth as polished granite, encased in white and silver suits like knights in Teutonic battle armor. Skylab's star dwellers by contrast slouched around in ocher-colored jumpsuits, as if they'd just been regurgitated by a giant pumpkin. They grew disreputable beards and gazed at the camera with sleepy squints, like roadies for Foghat on a West Coast tour. In the days of Gemini and Apollo, mission patches were dynamic little medallions filled with charging stallions and fierce eagles, vectors and digits and vaguely sexual penetrations. For Skylab, the insignia were static and oddly vegetative, featuring comforting sunbursts and holistic circles and a great green tree of life, like something you could get at a macramé booth at the Berkeley Farmers' Market Holiday Craft Fair. The three astronauts on Skylab 4 even staged what some in the press termed a brief mutiny—more like a "be-in," to use seventies terminology—to protest the crushing workload and unpleasant living conditions they had to deal with. This characterization, it should be noted, has been strenuously rejected by NASA and writers like space historian Emily Carney in the years since. Certainly it was a mild mutiny, if it was a mutiny at all. Nonetheless, it seems clear that there was tension between the astronauts and JSC's ground crew during the early stages of the mission.

The decay of Skylab's orbit—its gradual loss of altitude as the decade wore on—was inevitable, and indeed well documented. NASA wasn't worried at first, because the agency anticipated that something called the space shuttle would be online in time to travel to the space station and boost it back up into a sustainable trajectory. As late as 1978, in fact, there was talk of having the shuttle visit and reactivate the station for renewed habitation. It never happened. Delays in production of the shuttle dragged on, and unusually high solar activity heated the Earth's atmosphere just enough to slow Skylab's orbit, which in turn increased its rate of orbital decay. The crash of a nuclear-powered Soviet satellite in northern Canada in January 1978 gave people around the world a preview of what was coming. The resulting low-level panic saw entrepreneurs selling Skylab Repellent, while some residents of England hid in a cave, and the *San Francisco Examiner* offered a prize for finding the biggest piece of the fallen satellite. America's reputation for engineering excellence hit a new low when a man was reportedly killed in Indonesia after he offered another man the ultimate insult, calling him a Skylab. The errant space station entered Earth's atmosphere on July 11, 1979, breaking up a mere ten miles above the planet's surface—much lower than predicted. Pieces of the seventy-seven-ton station didn't hit anyone, but property damage and risk to life and limb were certainly possible when the descent occurred and debris rained down on Western Australia.

So, Skylab: our little piece of Appalachia in the sky. By and large, the life of the nation's first space station failed to produce America's most memorable moments in space. But it was still spaceflight—daring and risky and always demanding. Time has allowed us a heightened appreciation of Skylab's achievements, quirks and all. Perhaps the program's most important accomplishment was its creation of a whole new field of study for NASA. This occurred when the astronauts of Skylab simply turned their telescopic lenses in the other direction—back, that is, toward the solar-powered oasis of Earth. If their lenses had been strong enough, they might have seen a slight, blue-eyed individual standing in an El Lago street, staring right back at them. His hair was silvered at the temples now, but he stood just as straight as he always had. It was Bruce McCandless, disappointed but not despondent, waiting for his chance to touch the other side of the sky.

10

THE TEMPORARY
ANARCHIST

A lot of what I remember about growing up near Houston in the seventies will be familiar to the million or so other kids who did the same thing. Wrestling impresario Paul Boesch was a World War II hero whose enormous ears appeared to have been made from mashed potatoes and pressed onto his head by a drunken fourth grader. On Sunday mornings he hosted televised replays of Houston Wrestling's Friday night fights. It was my first exposure to participatory fraud, and I watched the slapstick ballet with equal parts delight and suspicion as local favorite Wahoo McDaniel regularly rose from the ranks of the semiconscious like some dime-store messiah to claim whatever bogus jewel-encrusted title belt was in play that week. When I was a middle schooler, birthdays meant a trip to Astroworld or ice skating at the Galleria and a side trip to Farrell's Ice Cream Parlour, home of the Zoo, a sordid carnival of lumpy ice cream and iridescent sauces that filled a platter big enough to bathe a baby in. At night I watched the news on Channel 13, which featured the breathless reporting of Marvin Zindler, a flamboyant former sheriff-turned-investigative journalist who combined the worst tendencies of Eliot Ness and Liberace. When he couldn't find anything better to report on, Zindler recapped the Health Department's report cards for local restaurants. "SLIME in the ICE MACHINE!" he'd shout, as if the slime was an alien life-form about to devour us all.

I went through the first five grades of school at El Lago's Ed White Elementary, named for the fallen astronaut. My classmates and I were almost all the offspring of space center employees. We were diligent and dutiful, though occasionally mouthy, and many of us were excellent spellers. We figured we were well on our way to . . . well, wherever we wanted. We had no idea what was about to come next.

Starting in sixth grade, we boarded big yellow buses and crossed Highway 146 to get to Seabrook Intermediate School. Though S.I.S. was, like my elementary school, almost all white, the school drew from the blue-collar communities on the western edge of Galveston Bay—Seabrook, Kemah, Bacliff, and El Jardin. These kids were tougher than we were. They smoked. They cussed. They had strange facial hair, and they would kick your ass if you crossed them. Some of us El Lago residents had uncles and cousins who'd protested the Vietnam War. These kids had kin who had died in it. They were the children of shrimpers and roughnecks and pipe fitters, and they'd done things we only whispered about. They were as wild as swallows in rotting barns. I felt the pull of their reckless glamour and wanted every part of it. And I wasn't the only one.

Suddenly my friends and I believed in everything and nothing. We were aching for trouble. We were confident, sarcastic, filthy, all knowing. We had greasy hair and grimy jeans, and we walked in packs whenever possible. We scratched ourselves and choked each other and traded conspiracy theories on gritty gymnasium floors. We roamed area pastures looking for psychedelic mushrooms in the cow manure. We were casual anarchists. We stole donuts and pencils. We were fascinated by sexual rumor, by the power of the Ouija board, by ghosts and Nazis and high school epiphanies on orgiastic beaches. The world was inconceivably big and strange, but there were secret words you could learn to navigate it. We were *this* close, we believed, to learning them.

The worst of those years for me was 1974. I spent way too many school days hiding out in the woods, where my buddies and I built ramshackle forts, played with kerosene fires, and drank gin and orange juice until we puked. To this day I'm plagued by gaps in my education attributable to those days of serial truancy. I have no idea how to multiply fractions, for example, and consequently have had to employ professionals to do my fraction multiplying. I sometimes act like I know what a gerund is, but the fact is, I wouldn't know a gerund if it sprayed me with a deadly neurotoxin, which is not something I think gerunds actually do. My dad disliked my hair and my clothes, my lack of interest in personal hygiene and my terrible grades. We argued on a regular basis, but our mutual disgust was more often expressed by each of us pretending the other didn't exist.

In my defense, 1974 wasn't such a hot time to be an adolescent. It was the era of Spiro Agnew and the Sharpstown Scandal, Evel Knievel's failed Snake River jump and the flyby of the overhyped Comet Kohoutek. We'd lost Vietnam to a bunch of dudes in black pajamas, and everyone knew it. President Nixon was a beady-eyed paranoid bastard and everyone knew that too, though not everyone said it. And my little corner of Texas was lousy with drugs. Ken Kesey and his Merry Pranksters had passed through Houston more than once in the sixties, and you could still smell the exhaust of their psychedelically painted International Harvester bus. It was like the decade had ordered too many hallucinogens, and now we were living through this massive warehouse sale of excess inventory.

El Lago was just a few minutes away from the Johnson Space Center. It was the kind of subdivision where men smoked pipes and built fiberglass kayaks, and the thought of a *National Geographic* special made folks want to hurry home after work. It didn't matter. The high school kid from down the street who taught me to throw a screwball later died from a heroin overdose. Three of my Boy Scout troop buddies cooked up a plan to break into a Galveston sporting goods store and steal guns. They said they aimed to start a revolution, though the goals of the insurrection were unclear. One night they climbed onto the roof of the store and broke in through an air-conditioning duct. They were busy smashing display cases and loading up on weapons when the cops showed up. As you might imagine, I was deeply disappointed in my friends—mostly because they hadn't bothered to include me.

One of my buddies from neighboring Clear Creek High School found out that he'd been cut from the baseball team and that his parents were getting divorced, with both of these news items delivered in the same week. He later became a professor of anthropology at a college in the Pacific Northwest, but that night in February as we walked to Mr. Gatti's for all-you-can-eat pizza, he was just another kid trying to figure out what the hell was happening around him. "I already knew you don't always get what you want in life," he said. "But this is worse. Now I know you don't always get what you deserve." It was teenage angst, I guess. John Hughes did it better in his movies a decade later. But I still remember that conversation forty years after it happened.

The decade was unsettling for adults as well. There were strange currents in the culture, curdled remnants of the heady brew of rebellion and experimentation of previous years. The first no-fault divorce law was passed in California in 1969, and divorce rates across the nation rose steadily afterward. Old bonds were breaking. The parents of one of my best friends were propositioned to join a foursome by a couple that lived down the street. (They declined.) The mother of a family we knew in San Antonio had an affair with a teenage boy she gave piano lessons to and later bore his child, though she kept the truth of the baby's paternity from her husband. The book *Open Marriage*, a sort of road map for authorized infidelity, was published and stayed on the *New York Times* best-seller list for forty weeks, its popularity apparently unharmed by the fact that my mother refused to speak its name. Like my father, Bernice McCandless had her superstitions. Forget about mortality. Hers was morality. She'd have taken a monogamous death any day over the prospect of a prosperous polygamy.

Some of those kids I knew at S.I.S. remained outsiders, rebels to the end. In high school they gathered each morning at the barricades, a metal traffic barrier at the dead end of a road beside the high school, where they blasted Aerosmith and Led Zeppelin on their car radios and smoked whatever they could roll in those little yellow sheets of paper they carried. One of them, a defensive tackle on the football team, died a few years later when a load of steel pipe fell on him at a job site. Another died in the process of trying to kill his stepfather with a homemade pipe bomb. The best player on my Little League team, a slick-fielding shortstop, became an alcoholic and one day blew his brains out in his own backyard. But for most of us NASA offspring, those years of stems and seeds were just a walk on the wild side. The rebellion didn't last. By tenth grade we had quit smoking pot and cut our hair and begun to act like the future urologists and bankruptcy attorneys we were. College was waiting out there like some strange combination of giant toga party and an extended trip to the dentist. Graduate school would come next. A job. A mortgage. Sublimation in the great churning sea of American ambition.

Despite my outward conformity, though, things no longer looked the same. To a pimply teenager with aesthetic pretensions, Clear Lake was a purgatory of sawgrass and sunburn, a dream-dulling region without a mountain or a valley, a river or creek or any particular feature at all save the columns of cumulus clouds that passed overhead, city-size sky walruses wallowing their way north, drunk with the salty liquors of the Gulf. Everything was new, bright, and cheap. A highway overpass was our biggest hill. Texas was swarming with refugees from colder climes, and the Gulf Freeway was a perpetual construction project. To the south, Galveston was close, but the water was dirty and the horizon marred by the sight of drilling platforms squatting over the ocean. Houston to the north looked like some massive blender full of asphalt had split apart at high speed and scattered its contents for a hundred miles in every direction. Admittedly, downtown was impressive and grew more so in the seventies and early eighties, with the addition of such structures as One Shell Plaza and Pennzoil Place and, though technically "uptown," the iconic Transco Tower, a building straight out of a Batman movie. On a clear day you could see these spires of the business district from Clear Lake. It was inviting, an Emerald City of sorts, but to get to it you had to pass through miles of freeway clutter—pavement, billboards, car lots, and fast-food franchises—until you hit the inner ring of roads near the University of Houston and the astonishing acres of whitewashed shacks of the Third Ward, a reproach to the eye and understanding. And then ... *nothing*. In that era, downtown was a business district, period. There was no street life: no sidewalk cafés, winsome buskers, or shady parks. During the day, when office workers moved about in tunnels beneath the streets to avoid the subtropical heat, downtown seemed eerily empty. The district died after dark.

The older I got, the less I liked my surroundings. For me, the attempts to create a culture around the space center seemed gimmicky and obvious. Timber Cove had a community swimming pool shaped like a Gemini capsule. Clear Lake's version of a debutante ball was called the Lunar Rendezvous. The Colt .45s, Major League Baseball's only team named after a firearm, became the Astros, and my Pop Warner football team, the Comets, played against a rival squad called the Space Bandits. As a teenager I learned that my new high school's official song had been written by a local advertising firm. That seemed to sum things up. My hometown was humid, homogenous, and unremarkable. Like Holden Caulfield in a Hobie T-shirt, I despised it all.

You may wonder why I'm telling you all this, since it apparently has nothing to do with Bruce McCandless and his shiny rockets. For me, though, the stories are all tied together. Junior high and Skylab. My father's long wait for a flight and the nation's garish aspirational hangover. If this memoir can be considered a belated and ineffectual love letter to my parents, as imperfect and spectacular and contrary as they were, it starts with the fact that they kept themselves facing forward through a weird and difficult decade. I certainly didn't help with this task. As a thirteen-year-old proto-alcoholic, I think I made things a good deal harder. But unlike many of their peers, my parents stuck together. This involved love, of course, as variously defined, and a dozen different categories of duty. But sometimes I think it was attributable above all else to stubbornness. Marriage was an important sign of success to their generation. My parents didn't like losing.

Bernice McCandless was anxious at the best of times. When it came to my sister, she could be downright paranoid. This led to sleepless nights for all of us, because Tracy was an adventurous kid, and my mother's anxieties were infectious. Tracy was a born hippie—a slender, impish blonde with golden skin and a ready smile. She favored loud music and large crowds, and, unlike the rest of her family, she trusted strangers. She would talk to anyone. She liked Fleetwood Mac and Kiss and ZZ Top, driving to Galveston way past midnight, and smoking in the driveway at 3 a.m. while my mom spied on her through the curtains. One night Tracy took our Suburban out with a friend and managed to flatten all four tires, a feat I've only heard of being duplicated once. Another time she borrowed—"stole," my dad said—the car for a trip to the beach and brought it back with strange indentations all over the hood. It looked like a metal moonscape. It turned out she and her friends had been dancing on it. Tracy was our red bandanna. She was the peacock feather amidst our family's olive drab, a little fire within the walls of the fortress.

It turns out Mom was right to worry. The seventies were years of rootlessness, anonymity, and motorized mobility. There were predators out there. The space center was a hub of science, dedication, and intellectual excellence, but its gravitational field only extended so far. America stood just outside this sphere of influence, and it hadn't shaved in weeks. In August 1973 a man named Dean Corll was shot dead in his home, maybe ten miles north of El Lago, by a teenager named Elmer Wayne Henley. Henley called the police to report the shooting. Upon questioning, he confessed to helping Corll rape, torture, and kill at least twenty-eight boys and young men, at the time the highest body count of any serial killer in American history. Corll was depraved and increasingly vicious. Over the course of the following year details trickled out in the local papers about the ways he lured victims to his "torture board," where he raped, sodomized, and eventually strangled or shot them. Several of the victims were buried on High Island, on the east side of Galveston Bay.

Ronald Clark O'Bryan, known as the "man who killed Halloween" for poisoning one of his own children with holiday candy, lived in nearby Deer Park. A local petroleum engineer named Kerry Crocker made headlines for sickening one of his two sons with radioactive pellets, castrating and almost killing the boy, whose only offense was siding with his mother in an acrimonious custody dispute. Girls went missing with alarming frequency in the area. Just down the road were the so-called Killing Fields, a weedy swath of land along I-45 between Houston and Galveston where something like thirty bodies have been found over the years. Many of the victims were young women between the ages of twelve and twenty-five, last seen in one or another of the towns just a few miles from our home in El Lago. All or most of the murders are thought to be the work of a serial killer, or killers, who is still at large.

My mom did the best she could to keep us close. Sometimes this was literal. She once ran screaming down the beach at Galveston when she spotted seven-year-old Tracy being led to a waiting car by a middle-aged stranger. Most often it was symbolic. Bernice McCandless wanted things to be "nice," she said, by which she meant normal and reassuring: Christmas trees and family reunions, anniversary presents and flags on the Fourth of July.

Every fourth or fifth Sunday, for example, we would climb into my dad's Volvo, which would be parked in the carport. The Volvo was a charmless vehicle whose enhanced fuel efficiency was achieved by sucking elegance out of the air and into the carburetor. Still, it was marginally cooler than our Suburban, a vehicle capable of seating an entire children's choir. SUVs were a rarity in those days. The Suburban was so big and stood so high off the pavement, it felt like traveling in a low-flying blimp. The thing drew so many curious stares that I frequently lay down on the floor of one of the two back-row seats so no one would see me with my family. At any rate, we took the Volvo. Dad would release the parking brake, which he always engaged despite the fact that we lived on a coastal prairie and the vehicle had nowhere to roll, start the car, and twist at the hips to look out the rear window as he backed out of the driveway. His breath smelled of either toothpaste or peanut butter. He would be in a bad mood, like the rest of us, because none of us particularly liked going to church. But my mom sometimes despaired of our souls and managed to coerce us into the car. My dad would be late to get ready, which would trigger a round of recriminations. My mother would threaten not to go, my dad would then insist that everything was going to be fine and it was *time* to go, and my sister and I would load up for the lips-pursed fifteen-minute drive up Highway 146 to La Porte, where we attended, fitfully, St. John's, at the corner of South Broadway and East G Street. It was "low" Episcopalian, presided over by an imposing but avuncular minister named Leighton Younger, a former corporate lawyer and University of Texas football player with hands the size of oven mitts. Episcopalian hymns are ponderous and mostly written by a guy named George Montgomery Crumlish in 1853, before the invention of rhythm. This was bad enough, but to make it worse my dad would sing off-key because he was tone deaf. Occasionally someone would glance sidelong at us, as if wondering if my dad was joking, but Dad never joked in church and just kept singing, his voice exploring some weirdly contoured territory all its own until the hymn came to a moderately triumphant but muffled end somewhere around verse 36. Finally we'd stumble out of church into the sticky Gulf Coast heat with gratitude that the ordeal was over and our souls were sanctified for at least a week. Peace would reign in the McCandless household for an hour or so. Then it would be time to mow the lawn.

"Onward," Dad would exhort. And then: "What did you say?"

EVERYONE HAS TALES TO tell about that great weird wasteland of years when your parents seem like tyrannical frauds and your face is a moonscape and life is engaged in a massive conspiracy to make you feel like an idiot. It was no different in 1974 than it is now. Occasionally, though, you hear a grace note that makes up for some of the static. My war with my dad was at its worst the year when I went off to summer camp. My mother was in New Jersey, visiting her sister, so my father drew the assignment of getting me to East Texas for five weeks of shooting rifles and singing church songs, which doesn't make a lot of sense unless you grew up in the South.

Dad chose to send me by bus. We didn't speak a word the whole drive into Houston. He was disappointed by my American flag T-shirt, which he considered disrespectful, and my sneakers, which had a hole in the toe, and my attitude, which had a hole right in the middle. He remained silent as he retrieved my trunk from the backseat of the car and carried it to the Continental Trailways station. It was a blistering hot day in July, and the sidewalk was measled with black spots of flattened gum that would stick to your shoes if you weren't careful, so I had a good excuse to look down. Dad found my bus and handed my trunk to the driver for stowage in one of the big luggage bays. The driver was a stout African-American man whose skin was glistening in the heat. I handed over my duffel bag. Dad nodded.

"Okay," he said. "All set?"

"All set," I confirmed.

"I'll tell your mother to write."

"Okay."

With this elaborate exchange of affection and encouragement, we parted. I climbed onto the bus. My hair was so long it needed a separate seat, so I took two. I settled in next to the window and opened my book, Maxwell Maltz's *Psycho-Cybernetics*, which I'd plucked from my mother's shelves because it promised a surefire method to improve my self-image. I didn't mind being put on a bus, I told myself. I didn't care that I was leaving for five weeks and no one seemed to give a damn. I was vaguely aware of other passengers boarding the vehicle and seating themselves around me, but I tried to ignore them. At last I heard the pneumatic hiss of the brakes. The bus lurched forward and wheeled out of the station onto Fannin Street. Gears gnashed and complained as we picked up speed.

Something caught my eye.

I looked out the window. There was a crazy man loping alongside the bus, shouting as he ran. We locked eyes. *Weird*, I thought. It looked like my dad.

"Stop the bus!" someone called. "There's an emergency."

The bus shrieked to a halt. The driver opened the door.

Dad vaulted up into the passenger well.

"It's for my son," he panted, holding a ten-dollar bill. "I forgot to give him his camp money."

Several moments of silence ensued. At last the driver turned and scanned his vehicle. His annoyance was palpable.

"Well?" he said, his gaze meeting mine. "You want your camp money?"

I walked to the front of the bus in a haze of embarrassment and gratitude.

"Excellent," said Dad as he handed me the damp bank note. Then, to the driver: "Righto. Thank you for stopping. I didn't mean to slow you down."

"Oh, don't you worry about that. He got everything he needs now? Toothbrush? Toothpaste?"

My father wasn't good with sarcasm. He didn't use it often, and he didn't recognize it when it was deployed at his expense. "A brand new tube. We checked before we left."

"Hallelujah," said the driver.

Dad waved as we pulled away from the curb. I turned and watched as he gradually grew smaller. He was still waving.

"That your father?" said a passenger across the aisle from me.

I nodded.

"Man can *run*."

11

THE FORGOTTEN
ASTRONAUT

I n the early days of crewed space missions, every astronaut was qualified for every flight. It was like being on a football team consisting entirely of quarterbacks. Because only one (Mercury), two (Gemini), or three (Apollo, Skylab, and the Apollo-Soyuz one-off) astronauts could actually go up on any given mission, competition for a flight was fierce. It wasn't just what you could do that mattered. It was also important to know what you *couldn't* do. A number of factors could knock a man out of line for liftoff.

It didn't take something lurid and ridiculous (like one astronaut's alleged diaper-clad drive from Houston to Florida to kidnap and, according to local prosecutors, murder her lover's new mistress) to lose your place in the lineup. Deke Slayton, one of the fabled Original Seven, was disqualified from the early crewed missions due to a latent heart condition. John Bull, a Group 5 astronaut, resigned from the program after being diagnosed with pulmonary disease. Ken Mattingly had to relinquish his place on Apollo 13 because he'd been exposed—just *exposed*—to German measles. Duayne Graveline was summarily dismissed from the program when his wife filed for divorce, and Apollo 7 pilot Donn Eisele never flew again after he became the first astronaut to *seek* a divorce (though others who were close behind him were not similarly penalized). Scott Carpenter was blackballed for acting more like a scientist and not enough like an engineer aboard Aurora 7, and Dave Scott, Al Worden, and Jim Irwin were kicked to the curb after news of the Apollo 15 Postage Stamp Scandal—which involved carrying unauthorized postal covers in the spaceship—broke. It was a big deal in the early seventies, though little remembered today.

Slayton became head of the astronaut office—chief astronaut, the position was sometimes called—after he lost his flight status in 1962. He went on to hold other management positions, including director of flight crew operations, and was for many years responsible for assembling the crews for manned missions. He was at least being frank when he said he preferred test-pilot experience in assembling his teams. Slayton clearly had his favorites, chief among them his bear-hunting buddy, big-knuckled Gus Grissom, who might well have been the first man on the moon if he hadn't perished in the Apollo 1 fire. After Slayton handed over the business of astronaut selection in 1974, it was hard to tell *what* went into the flight-selection process—what lent a man that secret, sacred aura of righteous stuff that qualified him for launch. I imagined the Soviets probably had a simpler way of choosing crews: hand-to-hand combat in a circle of burnt-out Panzer tanks just south of Stalingrad, each man armed with his choice of a giant hammer or a razor-sharp sickle that gleamed in the firelight.

As a teenager, all I knew is that my dad, and hence the whole family, was being disrespected. I wondered why it was all so mysterious, so hush-hush. What made it worse was that complaining about not being selected was itself evidence that a man didn't have the right temperament to go into space. My dad never grumbled about it to us. He wouldn't even discuss it. And so it was that in 1976, as the Apollo and Skylab programs faded into memory, the McCandless family headed to Denver. We filled the Suburban with spare underwear and low-grade discontent. We couldn't agree on a radio station. And my dad, dispirited but determined, drove 55 mph the whole way, gripping the steering wheel with both hands and trying not to glance too frequently up at the sky.

I don't know why it took Bruce McCandless eighteen years to get a mission. Some observers have said it was because he wasn't a test pilot, others that it was because he was the youngest man in his group, or that he was more scientist than fighter jock in his outlook and interests. His background might not have helped. Several of the early astronauts grew up poor. Slayton claims to have been tied to a tree by his mother when he was a kid so he wouldn't wander off. Stuart Roosa went to school, his daughter writes, with holes in his shoes. Al Worden's grandfather would turn off the engine of his truck and let the vehicle coast the final quarter mile into town, in order to save gasoline. Others worked odd jobs as kids, saved their dimes for flying lessons, flew combat missions in World War II and Korea. My father, with his more cosmopolitan roots, his varsity letter in sailing for Navy, and his postgraduate studies at Stanford, was not a natural fit with the rough-hewn likes of Slayton and Grissom.

At one point Dad thought maybe he'd just gotten involved with the wrong projects. In 1975, for example, he said, "I got to working on experiments for what was then being called Apollo Applications, and I just pretty well stuck with that ever since. I feel this may have been a mistake in retrospect, and I suspect that if I had gotten involved in the more operational aspects of the Apollo program, well, I might have had a better chance of flying on a [lunar] mission."

I suspect his personality had something to do with the delay as well. He was a brilliant man, but he wasn't shy about demonstrating his gifts. He insisted on doing things himself. For example, one day when we were visiting the space center, the Volvo failed to start. He cranked the ignition several times to no avail. He checked under the hood: nothing doing. He then took off his jacket and crawled under the vehicle to see what was wrong. Bruce McCandless was a gentleman in most matters. I never heard him tell an ethnic joke, or denigrate a woman, or argue with a guest. But he was also a naval officer, and he believed, in keeping with the best traditions of the service, that manual labor was best performed while swearing profusely. Sure enough, I soon heard curses emanating from under the car. Someone stopped to help. I've never been quite sure who this individual was, but he was classic NASA: close-cropped hair, heavy on the Vitalis; black-rimmed glasses; pens in his shirt pocket. "Need a hand, Bruce?" he asked. "NO, goddammit!" my dad shouted. That was it. Nonplussed, the man wandered off. He could have been an intern who'd started as an under-clerk in the

Obscure Data Procurement and Processing Directorate that very day. On the other hand, he could have been flight director Gene Kranz. Dad never thought to ask—or even *look*. It was extraordinarily antisocial behavior.

My mom always attributed his long wait to the fact that he wasn't a backslapper—that he lacked the glad-handing, self-deprecating gene that helps in any arena where males compete while professing camaraderie. Bruce McCandless was immune to most of the types of social intimidation we practice on each other. Al Shepard's Icy Commander stare? No effect. Deke Slayton's let's-blow-the-bridges growl? Didn't work. This could be helpful. He wasn't afraid to speak his mind, and—in engineering matters— he was usually right. On the other hand, his lack of emotional intelligence blinded him to veiled invitations and informal alliances, left him unable to detect the innuendo and intimation helpful in navigating any big organization. He realized at some point that if he was ever going to travel into space, he would have to make his way himself: work harder, press—gracelessly, if need be—on his projects, get things done exactly right.

Then too, my father was independent. He didn't like following orders for the sole sake of conformity. There's a single reference to Bruce McCandless in Walter Cunningham's big book about the space program, *The All-American Boys*. Cunningham is wondering why some of his fellow astronauts are getting good assignments while others aren't. He looks for clues in even minor behavioral characteristics. Why, he wonders, won't Al Bean fly at night? Is he afraid to? And why won't Bruce McCandless fly in *formation* like everyone else, goddammit? One of the astronaut wives put it a little differently. My dad, she said, "always seemed to be pulling his own wagon."

Astronaut James "Ox" Van Hoften remembered Bruce McCandless as "a very unusual character . . . [He] was always inventing something and trying to get them to build it. So it was kind of fun back then." Another astronaut Dad worked with for years once theorized that so much of my father's brain was dedicated to processing power that he had little left over for interpersonal relationships. This was her polite way of saying that though he was smart, Bruce McCandless could also be abrupt, self-absorbed, and prickly. In other words, he was hard to get along with. In one interview, astronaut

Hoot Gibson said my dad could be "a little difficult, sometimes," before walking himself back a bit. If there is one abiding rule in the astronaut office, other than *There's No Such Thing as a Bad Space Flight*, it's *Never Criticize a Fellow Astronaut*. Noted Gibson, "I shouldn't say *difficult*. With Bruce there was one way to do it, and that was the right way."

Another of my mom's hypotheses was that my dad was simply too brainy—or possibly too nerdy—for his own good. The other astronauts were jealous, she said, because he was *smarter* than they were. I find this unlikely, given the intellects and aptitudes involved in the space program, but I've always admired my mom's blatant partisanship in these matters. On the other hand, there is some testimonial evidence on this point. According to a 1984 article in *The Washington Post*, Dad was the first astronaut to wear a calculator on his belt (presumably Hewlett-Packard's HP-35, which was issued to Skylab astronauts to replace the slide rules used by Apollo spacefarers), which some of his colleagues dismissed as an intellectual affectation. Apollo 12 moonwalker Al Bean, who eventually became head of the astronaut office, was once asked why Bruce McCandless had to wait so long for a flight. One can almost see him scratching his head, searching for the polite answer. "The thing I remember most about Bruce," he said, "is that he was brighter and more capable than the rest of us. That's why he waited so long . . . In a competitive crowd like the astronaut corps, it's sometimes better to be behind the group instead of ahead of it like Bruce always was."

Bean's answer is a compliment, but it's not a very convincing one. There were lots of bright and capable astronauts. In fact, *bright and capable* is sort of what it means to *be* an astronaut. Nor did the spacefarers have a monopoly on brainpower; NASA and its contractors were the nation's largest employers of hyper-intelligent engineers, technicians, and managers. But the diplomacy of the response doesn't obscure the truth beneath it. Whether you're behind the group or ahead of it, you're *apart* from it, and that seems to have been the case, for whatever reason, for Bruce McCandless. In photographs, especially when he was young, he typically stands or sits at the edge of the group—not far distant, but certainly not in the middle of it either. In a group shot taken on a geology field trip to Iceland, most of the astronauts lounge around in military gear and sneakers, athletic men all, at ease in their superstar status and animal spirits and casually dismissive of the photographer addressing them, the whole unruly bunch ready and frankly eager to get out of this damned volcanic wasteland and into a martini or two.

My dad, by contrast, stands erect, almost at attention, just behind the group, wearing a scarlet shirt and what appear to be bedroom slippers. He seems to have missed the memo authorizing use of the American Slouch, Piercing Gaze variety, until further notice. He's ready to go out and collect shale samples, by God, and I can attest that the enthusiasm is genuine. I found rocks from his various expeditions still wrapped in newspaper among my father's files after he died. I suspect he was the most ardent geologist of the bunch, bedroom slippers or no. Jack Lousma said that on geology expeditions, men in the group would sit down to lunch and wonder aloud, "Where's Bruce?" No one knew. "So we'd go look for him," said Lousma, "and he'd be sitting by himself on the next ridge over, contemplating something interesting he'd seen. He was a true scientist."

Even more unlikely than her hypothesis about Dad's envy-inducing brain, Mom blamed *herself*. She wasn't a good astronaut wife, she speculated. People didn't like her. She wasn't pretty or witty or social enough. My mother was beloved by just about everyone who ever met her, so this seems highly doubtful. But who knows? In his book *Riding Rockets*, Mike Mullane writes caustically about the utter inscrutability of the flight-crew-selection process. Walt Cunningham discusses it at length in *All-American Boys*, while conceding in the end that he was never quite sure how the process worked—even when, early on, it favored him with an assignment to Apollo 7. Astronaut Kathy Sullivan, the first woman to walk in space, comments that "the reasoning behind the particular technical assignment each of us got [when starting at NASA] was a complete mystery to us, as would be the logic behind every other assignment in our astronaut careers, along with the process by which the decision was made."

While my mother, my sister, and I were thinking black thoughts about NASA and the astronaut corps as a monolithic entity conspiring to deprive Bruce McCandless II of his shot at space, the truth was more complicated.

Matthew Hersch, in *The Making of the American Astronaut*, describes the astronaut corps not as a neatly structured group but as an agglomeration in which individuals competed for approval from management, and various groups jockeyed for position against the others: veterans vs. rookies, pilot-astronauts vs. science-astronauts, one class against the next in a slow-motion free-for-all in flight suits to determine who got a seat on the limited number of missions. It was a grueling existence. It wasn't just the competition. It was the fact that the competition was so *good*. In Building 4 every one of your peers was Captain Kirk. Dad knew it, too. Bruce McCandless occasionally rolled his eyes at the bureaucratic excesses of NASA as an organization, but I never heard him say a negative word about his colleagues. He was a fan. Paradoxically, it was the fierceness of his desire to be worthy of inclusion in the group that made him so standoffish. He wasn't going to ask anyone for anything. He was going to be the hardest working, most self-reliant man at JSC.

That's one theory, anyway.

Rick Houston and Milt Heflin suggest in their book *Go, Flight!* that my dad's refusal to follow instructions from Deke Slayton while working as a capcom during Apollo 11 may have had a much more tangible effect on his flight prospects than any general personality trait. My dad eventually heard the rumor that he'd offended Slayton that day in July 1969 by not heeding the instruction to bring the astronauts back into the lunar module. "Thirty years after the fact," Dad comments in the book, "somebody said that they'd heard [Director of Flight Operations Chris] Kraft remark the reason I didn't get a flight in Apollo was because of insubordination. The only incident I can relate back to was that one where I basically ignored Deke." My dad figured Slayton's request probably came from higher up—possibly even from Kraft himself, who was notoriously sensitive to any challenges to the chain of command. Dad admitted he had no idea if this was the real reason for his subsequent time in what he called the penalty box. Still, he didn't dismiss the possibility. In fact, toward the end of his life, he seems to have accepted it.

I frankly doubted the tale when I first heard it. It seemed too petty and punitive to be true, given the caliber of the men involved and the importance of the event. Bruce McCandless was simply doing his job that day. But if ignoring Slayton and Kraft was in fact the reason my dad fell from grace after Apollo 11, it's a perfect example of his identification with his fellow astronauts over the suits back on Earth, and a costly lesson in the price of obeying the rules over the rulers. Bucking authority may have cost him his shot at the moon. Something did, and I suspect it was his gradual realization that he'd screwed up—said the wrong thing, offended the wrong person—that gnawed at him over the next decade and a half. That left him staring moodily down at his plate at our dinner table as we, his family, wondered what we'd done wrong, and—more importantly—how we could fix it. It turned out, of course, that we *couldn't*. He was going to have to do that himself.

Could he do it? Of course he could. He was the smartest man on earth. It's just that he didn't always connect with people the right way. That Christmas he gave my mother a beautiful gold ring. Fashioned from metals salvaged from a Spanish galleon that had sunk in the Gulf of Mexico hundreds of years ago, it glowed like warm butter. She loved it. My sister, Tracy, and I traded glances. We were both happy that Mom and Dad were being kind to each other. But it lasted all of six minutes. Dad had wrapped one more present, it turned out. It was the last one Mom opened. She shucked off the paper, glanced at the gift, and seemed to shrink a size or two. It was a book.

"*Open Marriage*," she said. The words hung there like the air itself had been branded.

Tracy and I weren't the sharpest knives in the drawer, but we knew the significance of what my mom was now holding like a dead rat. It's possible my father had forgotten what was under the tree. Maybe he hadn't expected her to say the words out loud. Or maybe he just hadn't thought much about the optics at all. He took a quick read of the room.

Zero degrees.

Kelvin.

"Just for consideration," he offered.

Christmas dinner was late that day, and I'm pretty sure we skipped dessert.

12

THE LOON

It was the loon that gave it away. Before then I'd thought maybe everyone's dad fed random opossums and brought deadly snakes home in burlap bags. But Bruce McCandless's care for an injured loon during the summer of 1975 clued me in that his commitment to the cause of conservation ran deeper. In fact, my father was the first environmentalist I ever knew: an energy-conserving, grebe-loving, no-hunting acolyte of ecological consciousness sitting across the breakfast nook from me.

It probably wasn't something he wanted to advertise. Being an environmentalist in Houston in the seventies was a little like being a Christian in Imperial Rome. Home to major operations of Texaco, Shell, and Exxon, the rapidly expanding city had taken to calling itself the energy capital of the world. Houston sweated crude and belched natural gas. Houston hurried up and down its freeways under a haze of smog and liked it that way. In this empire of steel and glass, environmentalists were heretics, outliers, odd and possibly dangerous, except that there were so few of them that they were really more of a nuisance than a threat. And yet, to use the modern phrase, they persisted.

I know of two important influences on Bruce McCandless's views. In December 1968 the crew of Apollo 8 brought back those famous Earthrise photographs, shots of "a crescent Earth rising above a lunar horizon against an inky black sky." *The Whole Earth Catalog*, an analog Amazon for the counterculture, claimed that these images "established our planetary facthood and beauty and rareness . . . and began to bend human consciousness." Apollo 17's photographs of the Earth in a single frame, the famous "blue marble" suspended alone in the sea of space, had a similar effect. The pictures showed my dad and many others how finite, indeed *small*, our home planet is, and he noted this impression in speeches he gave from that point on.

Another important influence was Rachel Carson's *Silent Spring*, which he kept on a bookshelf in his study. In the book, first published in 1962, Carson wrote eloquently about the deleterious effects of DDT in the food chain of North American birds. DDT was a potent poison used to control crop-eating insects. Some observers called it the atomic bomb of pesticides. Inevitably, the toxin washed into the nation's waterways, where the substance was ingested by small creatures that were in turn devoured by birds. Traces of the chemical in the bloodstreams of birds contributed to a thinning of the shells of their eggs, which led in turn to premature hatchings and the death of many of the young. Some of North America's most majestic birds, including the bald eagle, faced extinction. Estimated to number 100,000 in the late eighteenth century, the bald eagle population had diminished to fewer than 500 mating pairs by the early sixties. In response to such ominous developments, my father banned the use of pesticides or herbicides in or near our house. This may sound like an airy gesture, but we lived on the Texas Gulf Coast. It was the Cockroach Riviera, and the result of Dad's insecticide boycott was a life replete with roaches, fleas, raccoons, and rats. Though Bernice McCandless supported the ban in theory, she occasionally lapsed into profanity and despair when considering some of its multilegged consequences.

Pesticides weren't the only poisons. From California's smog-filled skies to Cleveland's burning Cuyahoga River, it was obvious we Americans were fouling our nest in a variety of creative and catastrophic ways. My father preached early and often against the common practice of dumping used oil and antifreeze into the neighborhood's storm sewers, which drained into Galveston Bay. He once stopped by the side of I-10 on the way back to Houston from Beaumont so he could photograph a refinery belching toxins into the September sky. He planned to report the factory owners for air pollution, he said—and I suspect he did, though on the Gulf Coast in those days it's unclear whether anyone cared. Galveston County has four EPA Superfund sites in various states of remediation or monitoring, while Harris County, where we lived, has twenty-three, including sites known as the Highlands Acid Pits and the San Jacinto River Waste Pits. A trip to Galveston in those days meant swimming with tar balls—floating globs of oil from drilling rigs, tanker ships, and, yes, natural seeps, that ended up on the beach and then on beachgoers' feet. Residents counseled newcomers not to be alarmed. That odd, slightly sulfurous odor coming off the Gulf? That was the smell of money. What was good for Exxon, it seemed, was good for the country.

Other spurs to Bruce McCandless's environmental commitment are harder to trace. He and my mother simply shared an enduring fascination with wild things, which is not what you might expect from a couple of kids from Long Beach and the environs of Newark, respectively. Jack Lousma recalled that my father often went off by himself on geology expeditions. Fred Haise, of Apollo 13 fame, remembers that on survival trips the astronauts took in the late sixties, my dad would wander off from the group while searching for "critters," as Haise calls them. "Where's Bruce?" was a common refrain, along with, as the hours wore on, "Where the *hell* is Bruce?" Because no one could get back to civilization if my dad was AWOL, his colleagues eventually came up with the idea of appointing a "Bruce watcher" to keep him from disappearing in pursuit of some unusual specimen.

I always figured such pursuits were spur-of-the-moment things. In my dad's files, though, I found a record of a call he made in 1967 to the Houston Zoo. In it he inquired, shortly before departing for training in Panama, about what sorts of animals the zoo might be interested in adding. An assistant manager named Dooley replied that the zoo wanted marmosets, howler monkeys, capybaras, small anteaters, and "any reptiles or vipers." The zoo did *not* want ocelots, coatimundis, or kinkajous. Toucans, said the notes, were OK. During the trip, Dad acquired a three-foot-long specimen of *Bothrops asper*, commonly known as the fer-de-lance. One of the most dangerous snakes in the world, the fer-de-lance is sometimes referred to as the ultimate pit viper, a creature whose bite can kill a strong man in hours. Despite the obvious discomfort of his fellow astronauts, Dad stuffed the serpent in a burlap bag and brought it back to Houston, where he personally presented the ultimate pit viper to the grateful but presumably wary Mr. Dooley.

THOUGH HE WAS FASCINATED by snakes and small mammals, Bruce McCandless's special passion was birds. It started early. In 1945 his grandparents sent him a copy of John James Audubon's *Birds of North America* for Christmas. In 1947 the ten-year-old wrote to his mother from camp that "your letter arrived yesterday, along with my Audubon Bird Leaflets. I have found some interesting facts about each bird that I ordered." Two years later the now twelve-year-old Bruce wrote home that he was "safe at camp," that he weighed seventy-four pounds, and that on the drive up he and his father had seen "all [kinds] of birds," including a nighthawk and a great blue heron. As an astronaut, Dad was made "bird control officer" for the Shuttle Landing Facility at the Kennedy Space Center. Whether this was an official post or just something NASA brass invented to keep him busy is unclear. Either way, he tackled the assignment with his usual vigor. Dad joked in a letter to his brother Douglas, "Will whoever has any ideas on how to discourage flocks of 20,000 or so wintering tree swallows from gathering on the runway please step forward?"

Evidently he had a few ideas himself. He studied the nesting, feeding, and breeding patterns of the egrets, herons, and, yes, swallows that live near the NASA landing strip at Cape Canaveral. He then briefed his astronaut colleagues on this information in an attempt to help them avoid encounters with the creatures, like the in-flight collision with a goose that had resulted in the fatal crash of astronaut Ted Freeman near Houston in

October 1964. The astronauts took the suggestions, said Apollo veteran Al
Bean, and never had an accident.

I remember the moment I realized that birds had gone from passion to
obsession for my father. It was on that road trip from Houston to Florida
to see the launch of Apollo 17. For much of the journey, the family was
treated to *A Field Guide to Western Bird Songs*, a series of cassette tapes
containing the characteristic calls of common North American species.
Because my dad was the stereotypical drive-till-the-wheels-melt alpha
male, the caws and croaks of jays, herons, and gulls filled the Volvo long
into the night. It turned out the aim of this exercise was to help him excel
in the upcoming Freeport Christmas Bird Count, which took place annu-
ally just down the coast from our home. I was invited to participate more
than once. Remembering that horrible night journey through the bayous
of Southern Louisiana while listening, involuntarily, to the mating burps
of the glabrous hornbill, I resolutely declined.

My father's obsession with protecting birdlife was admirable,
of course. But it probably didn't do much for his prospects at NASA. Occasionally his concerns could seem grating, even when couched in humor. In
1979, for example, he gave one of his fellow astronauts an "award" to com-
memorate the astronaut's having killed a rare Attwater's Greater Prairie
Chicken while flying in one of the agency's T-38s. In doing so, my father
wrote, his colleague had "reduced the local population of the species from 4
to 3 and consequently helped the [prairie chicken] a finite distance further
along the path towards extinction." Funny? Sort of. Appreciated? Not likely.

Dad eventually made himself into a wildlife-rehabilitation expert. In
the mid-seventies, he tended, treated, and returned to the wild a succession
of injured or orphaned birds, including three baby screech owls, wide-eyed
and ponderous; a red-shouldered hawk that was too powerful for us kids
to handle; a pied-billed grebe; and the common loon. He kept notes and

A young Bruce McCandless II with sundial (left) and father Bruce I (right).

Newly minted Commander Bruce McCandless addresses newshounds in San Francisco in December of 1942. His wife Sue stands at far left, next to his mother, Velma. US Navy photograph.

Bruce McCandless II with fellow Top 3 graduates of the United States Naval Academy Class of 1958, Alan Chodorow (left) and John Poindexter (center). US Navy photograph.

20-year-old Bernice Doyle in 1957.

Bruce McCandless II in May 1958, shortly before graduating from the Naval Academy.
US Navy photograph.

Soon-to-be Ensign Bruce McCandless II receives his diploma from
President Dwight D. Eisenhower, June 4, 1958. US Navy photograph.

Bruce McCandless II and author,
September 1961.

Wedding of Bruce McCandless II and Bernice Doyle,
August 6, 1960.

Officers of VF-102, the Diamondbacks, aboard the USS *Forrestal* in April 1961.
Dad is third from the left on the bottom row. US Navy photograph.

Lieutenant Bruce McCandless II entertains crooner Bing Crosby aboard the USS *Forrestal*,
September 2, 1962. McCandless was periodically sealed up in an aviation pressure suit and
displayed as an astronaut for visitors to the ship. Note the "squawk box" hanging
in front of the suit, and the oxygen tank in the lower right-hand corner of the picture.
US Navy photograph.

Group 5 astronauts meet the press in Houston in April 1966. Bruce McCandless II is fourth from the left on the bottom row. NASA photograph.

Survival training in Panama, 1967. From left, astronauts Gerald Carr, Al Worden, and Bruce McCandless II, training officer Raymond Zedeker, and astronaut Don Lind listen to a briefing from Air Force Major Jerry Hawkins. NASA photograph.

Bruce McCandless II on capcom duty
for Apollo 11 in July 1969. Beside him is
astronaut Jim Lovell, who would go on to
command the ill-fated Apollo 13. NASA photograph.

Bruce and Bernice McCandless at
Cape Canaveral before the launch of
Apollo 11. Photograph by Mike Geehan,
courtesy of Elizabeth "Betty" Geehan.

Astronaut geology field trip to Iceland, 1967. Bruce McCandless II is in the red shirt to the right and
rear of the group. NASA photograph.

Tracy and I joined Dad in a Skylab mock-up at JSC in 1970.
NASA photograph.

Bruce McCandless II and Deke Slayton (standing, center) watch simulated lunar exploration at the
Manned Spacecraft Center in early 1970. NASA photograph.

Bruce McCandless II tests Skylab
dining options, 1971. NASA photograph.

Bruce McCandless II tests the AMU in 1971. The device,
which evolved into the Manned Maneuvering Unit,
grew progressively bigger and heavier over the years
before being reimagined and produced, in 1994,
as the smaller and lighter SAFER. NASA photograph.

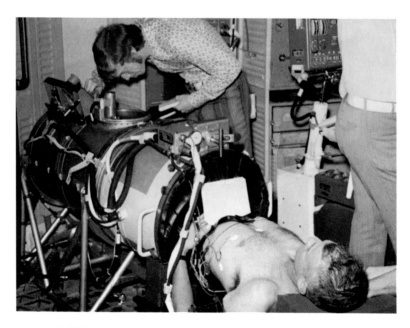

In 1972, SL2 backup crew members Rusty Schweickart (standing) and
Bruce McCandless II practice using the Lower Body Negative Pressure Device. NASA photograph.

Bruce McCandless II takes a break from Skylab capcom duty, 1973. NASA photograph.

Bruce and Bernice McCandless in JSC's Mission Control Center, 1974. NASA photograph.

Bruce and Bernice McCandless, Halloween 1977. Dad was a *Star Wars* fan.
After seeing the first movie in the series, he made himself a Jawa costume, complete with cowl,
a black scrim to hide his face, and glowing electric eyes.

Dad in 1975 with one of the three fledgling
screech owls he and my mother rescued,
rehabilitated, and eventually returned to the wild.

The McCandless family in
September 1983.

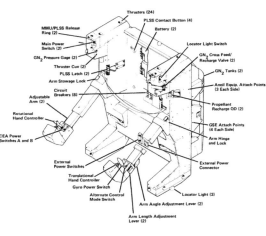

MMU Diagram.

In a 1972 NASA photograph, Bruce McCandless II (in pressure suit) and Ed Whitsett test a mock-up
of the Automatically Stabilized Maneuvering Unit in preparation for Skylab's M509 experiment.
McCandless is brandishing a version of the Hand-Held Maneuvering Unit, which was
fueled by a nitrogen tank in the ASMU. On the right is a diagram of the Manned Maneuvering Unit,
courtesy of Lockheed Martin Corporation

The crew of STS-41B steel themselves for launch in 1984. From left, the astronauts are
Commander Vance Brand, pilot Hoot Gibson, and mission specialists Bruce McCandless II, Ron McNair,
and Bob Stewart. NASA photograph.

In Hoot Gibson's famous photograph, Bruce McCandless II operates
the Manned Maneuvering Unit 178 miles above Earth on February 7, 1984. NASA photograph.

Riding the "cherry picker," courtesy of Ron McNair
and the Canadian-made robotic arm.
NASA photograph.

STS-41B heads for low-Earth orbit on
February 3, 1984. NASA Photograph.

Bruce McCandless II took this photograph of *Challenger* while he was operating the MMU. Bob Stewart can be seen on the right side (rear) of the payload bay. NASA photograph.

STS-41B: *Challenger's* seal-like nose and forward windshields are reflected in Bruce McCandless II's gold-tinted helmet visor as he flies Manned Maneuvering Unit No. 3 in February 1984. NASA photograph.

STS-31's deployment of the Hubble Space Telescope took place on April 25, 1990. NASA photograph.

The crew of STS-31, the Hubble Space Telescope deployment mission. From left:
Steve Hawley, Kathryn D. Sullivan, Bruce McCandless II, Charlie Bolden, and Loren Shriver.
Sullivan's book, *Handprints on Hubble: An Astronaut's Story of Invention*, describes the
laborious work that preceded deployment to make sure that Hubble could be repaired in space.
It was time well spent, as several repair and maintenance missions later proved. NASA photograph.

STS-31 astronauts Loren Shriver (commander), Charlie Bolden (pilot),
and Bruce McCandless II (mission specialist) review flight plans in April 1990. NASA photograph.

Bruce and Bernice with family, Christmas 2011. The kids in the front row are grandchildren Emma and Carson. In the back row, left to right, are Bruce III, wife Pati, Dad's sister Rosemary Van Linde McCandless, Bernice McCandless, and Bruce II. Photograph by Tracy McCandless.

Bruce and Bernice McCandless hiking on the shoreline of Washington State, 2011.

Bruce and Ellen Shields McCandless in Denver for the birthday of Ellen's son Steve Shields in 2016. Photo courtesy of Ellen Shields McCandless.

The Hubble Space Telescope caught this image of M16, the Eagle Nebula, in 2015. The "fingers" of this celestial hand are composed of gas and dust and are several light years long. Some astronomers say that a nearby cloud of hot dust, theorized to be the shock wave from a supernova, may already have destroyed this formation, and that we should be able to witness this destruction from Earth in about a thousand years. NASA photograph.

Dad took this photograph of the "Great American Eclipse" in August 2017, after traveling to Nebraska to witness the event in its path of totality. It may have been the last picture he ever took.

photographs related to these projects in a bound red volume. Some of the notes read like a true-crime procedural:

Wed. 5/28/75—1530 – Bernice telephoned with info that [name unclear] in Clear Lake Forest had an adult and three young screech owls . . .

1630—Clear Lake Forest . . . Adult was in bad shape; did very little except lie there and shudder occasionally . . . Other adult (parent) was dead on front lawn . . . apparently a day or so. Suspected acute pesticide poisoning, both adults . . .

1800—[Veterinarian] Hoover suspected organo-phosphate poisoning of adult—pointed out oscillating behavior of pupils of eyes.

The adult owl died the next day. My dad collected the fledglings, though, and he and my mother cared for them for the next three months. Dad fed them mealworms and shreds of raw beef when they were young and switched to using whole mice as they matured. He bought a toy loco-motive, set up a track, and placed frozen mice in the coal car. At night he would set the mouse train running around the enclosure, simulating the movement of prey in the wild. He used nail polish to paint the birds' tiny talons—designating each bird with a different color, red, blue, or green—so he could track their weight gain as they grew. He finally released them on August 22, on a night Dad described in the book as *full moon, no wind, very quiet, sort of misty—had been raining heavily in the afternoon.*

The owl project lasted longest, but some of my dad's other rehabilitation efforts were equally vigorous—and ingenious. For an injured nighthawk, he fashioned a spring-loaded traction system, using a snap swivel, a tubular tangle guard, pulleys, and a handmade bridle designed to keep the injured bird from falling over. The Red Book contains Dad's schematic of the device, with the bird ("Bird") helpfully labeled. He created a mini-estuary in our garage for the loon, *Gavia immer*, complete with wading pool, heat lamp, sandy beach area, and water circulating system, and fed the creature

tiny fish that we hauled out of Taylor Lake in a seine net each evening. It
was a lot of work, but the reward was watching the animal slowly recover.
She was an immature female, intense and skittish and as light as a bag
of sticks. On warm nights in July, we laughed as we heard that haunting
North Lakes cry coming from our garage. The loon had a serious gash in
her breast. She battled infection all summer, but by late September she was
strong enough for my father to release her in an inlet of Galveston Bay.

We kept turtles and tropical fish and fostered a raccoon that ended up
dying of distemper. It got to where neighbors declined to kill snakes in our
seven-street area without first checking with my dad. If the serpent was
nonvenomous, he would capture it, stick it in a sack, and free it somewhere
deep in the woods. That's how we got the eastern hognose, *Heterodon plati-
rhinos*, a mid-size reptile with a snout like a wood chisel. It was found by a
neighbor just down the street, who said he'd never seen a snake of its like
before. Indeed, the little hognose was so odd and interesting that we kept
the snake a year before rewilding it.

DAD'S ENVIRONMENTAL IMPULSES CHANGED him. Sometime in
the early seventies, he started wearing a coded message on his belt. It was
a circular brass buckle, die-cut to show a rising sun with rays of yellow,
orange, and green, and, above it, multicolored clouds. At least they *looked*
like clouds. Further inspection revealed that the clouds were actually styl-
ized letters spelling out the word PEACE. It was like looking at one of
those black-and-white drawings where the first thing you see is a goblet,
and after a while you notice two human faces in profile instead. I must have
seen that buckle fifty times before I deciphered the message. I'm looking
at it now, in fact, as I write. When I flip the buckle over, it says: ECOL-
OGY DESIGNS 1971, PEACE WITH NATURE. Needless to say,
such messages were a break from the image of the steely-eyed missile man
that guided NASA through its first three manned spaceflight programs.

The environmental movement affected the rest of us, too. At one point
or another, my mom, Tracy, and I were all expected to contribute to the conservationist cause. I fed the birds and fish when Dad was out of town, and
I paddled a canoe when we brought visitors on tours of the nearby Armand
Bayou Nature Center, which was named after the murdered Houston naturalist Armand Yramategui. My mother's environmental impulses were
less programmatic than Dad's, but they were equally intense. She and my
father shared a weakness for injured, abandoned, or otherwise unfortunate
animals, and did what they could to care for them. Because she was busy
raising two kids, Mom generally played backup on the animal-care front as
my father got more involved. He joined the Environmental Defense Fund
in 1972. An early member of the Houston Audubon Society, Dad eventually served as president of the organization from 1975 to 1976 and then
again from 1977 to 1979.

Houston biologist Fred Collins remembers meeting Bruce McCandless
in 1971. A local naturalist named Jerry Smith was working to create a data-
base of fish-eating birds that today is known as the colonial waterbirds
survey. As Collins, now the director of the Kleb Woods Nature Preserve in
Tomball, Texas, recalls:

> Jerry was having trouble finding all the colonies in and around Galves-
> ton Bay due to lack of access to and, in some cases, a lack of knowledge
> of where some species might be nesting. His biggest nemesis was the
> Forster's tern, which he knew nested in marshes. However, finding the
> colonies by boat or wading through marshes was woefully inadequate.
>
> About that same time, I was finishing up graduate school at Texas
> A&M and planning to move back to Houston. Jerry had asked the
> renowned Clarence Cottam, Director of Welder Wildlife Foundation
> at the time, for a young biologist volunteer to help with the survey
> efforts. With Dr. Cottam's recommendation, Jerry contacted me and
> asked me to help with the surveys. I agreed. On our second field outing,
> he had me meet him at Hobby Airport where I met Bruce. We boarded
> his seaplane and flew to Galveston Bay. It was my first flight with a jet

fighter pilot. I found out the reason he had purchased the plane was because it had the most horsepower of any small standard-manufacture single-engine private plane. It was capable of acrobatic maneuvers. I was not much of a flyer, being prone to air sickness. By the time we were in West Galveston Bay, near Snake Island, I was green and oh-so-happy to feel the bay beneath the plane when we landed. Snake Island felt wonderful to walk on. I may have actually lay down to steady my head.

But soon we were up again. The wings of the seaplane were below the cockpit, which had a jet fighter–like canopy, so seeing the ground required rocking and rolling. Flying down the back side of Galveston Island at about 300 feet with the plane alternating from side to side in a series of half rolls, I lost my lunch and everything else in my stomach, but Bruce found Jerry's missing Forster's tern colonies. Hundreds of birds, perhaps a thousand nests, were scattered on the points of the marsh grass that extended into the bay from various drainages and inlets. We were excited, and they wanted pictures! I just wanted to be back on the ground! It was just about impossible to take photographs as the plane rocked back and forth, but Bruce had a solution. We got to the end of the island and Bruce climbed straight up and over. We were upside down. He then flew back down the island upside down over the tern colonies and took pictures with his pistol-grip Hasselblad camera directly through the top of the canopy. In spite of my air sickness and disorientation, I was impressed, and I can still visualize Bruce doing this as if it were nothing unusual in the least.

Maybe the most visible project my father undertook with the society was establishing the Edith L. Moore Nature Sanctuary on the west side of town. I was roped into some of the sanctuary work as well. Not the legal work, of course, which led to the sanctuary's getting tax-exempt status, the first such organization to earn this designation under Texas property tax laws. I'm talking about the actual filthy *let's-pull-the-old-tires-out-of-Rummel Creek* work involved in clearing up the site. The sanctuary, a shady oasis in the midst of Houston's untold acres of sunbaked concrete and tinted glass, is now the Houston Audubon Society's headquarters. As an educational center, it serves some 10,000 Houstonians a year. As a sanctuary, it's quiet, relatively cool, and home to Texas flora and fauna including cypress trees, live oaks, snakes, turtles, and herons.

The Audubon Society was happy to have him. In the eyes of a world that looked at birders as odd and ineffectual, he was the opposite: a fighter pilot and astronaut, engineer and family man who was afraid of no one and knew how to use technology to make a point. And he loved what he was doing. Fred Collins remembers a spark in his eye when Bruce McCandless talked about environmental matters.

It should be noted that this talk wasn't particularly lofty. Bruce McCandless wasn't a poet. I suspect he felt the same way about words that I do about numbers. I realize numbers have a place in the world. I'm happy for them to fill that place in normal, sequentially predictive ways and then get on a bus and go back to wherever they came from. We need to know when it's halftime, after all. My father saw numbers as wondrous, flexible things, heralds of higher laws and spheres of wonder. But teasing the implied and ironic meanings out of a poem held no interest for him. He kept his metaphors under strict supervision. His sentences marched along in single file. Words were for work, and he kept them at it.

He was sensitive about the quality of his prose, though it always seemed intelligent and thorough to me. But it's true that he didn't speak or write about the Earth like a poet, or even like some of the other astronauts. He was an engineer. Even his raptures were regulated. Of the importance of environmental consciousness, for example, he once said:

> This leads me to the concept of Spaceship Earth, and I am not talking about the Spaceship Earth at Epcot, but about Spaceship Earth here. Now I am, or have been introduced as "The Astronaut," but everyone in this room is an astronaut, in the sense of being a crew member on Spaceship Earth. It's a very big spaceship, but it's not infinite, and yes it has systems. You don't open and close valves to drain hydraulic fluid or electricity, but you do have ocean currents, atmospheric circulation, we have concerns about pollution, ozone depletion, and somehow if we're

going to pass on a better spacecraft to our children and their generation, we need to manage the systems in their best interest.

That's not a *grito* that's going to make you want to run out and join Greenpeace. But it is true, and it's sincere. Bruce McCandless invoked the image of Spaceship Earth often. In a talk he gave to a group of Boy Scouts, for example, Dad noted that his generation was only borrowing the planet until the current crop of young people could take possession. Taking care of the spaceship was important because, regardless of what his audience might see in the movies, there wasn't going to be anyplace else for mankind to live for a long time. In a feature story on "The Environmentalists" published by *The Houston Post*, my father was even more modest. He commented that "I view environmentalism really as just responsible, concerned citizenship more than anything."

There may have been more lyrical advocates for the Earth and its systems than Bruce and Bernice McCandless, but there weren't many who cared more deeply than they did for its inhabitants. My dad got a rabies vaccination every year, and he and my mother did wildlife rehabilitation work their entire lives. In fact, one lesson I learned from Bruce McCandless is that dedication to space exploration doesn't imply a lack of interest in life down here. Curiosity is not an either/or issue. It's both/and. Space and Earth. Arcturus and Armand Bayou. When you look up on a clear night and marvel at the wheels of light in the sky and all the worlds that have to exist deep in the darkness, it's important to pause a minute and imagine looking the other way—from *out there*. Imagine gazing at Earth's paper-thin layer of atmosphere and its sapphire seas and white drifts of water vapor casting shadows on the globe. Which is the more amazing sight, you might ask, the blue or the black? And then wave at yourself, proud inheritor of this improbable world, and walk for a while under the whispering stars. Though he spent a significant amount of his life wanting to leave this planet, if only for short excursions, Bruce McCandless was always fascinated by Earth's many mysteries. There are, he sometimes said, wonders all around.

13

BUCK ROGERS
AND THE SILVER BIRD

I n his long leaden days of waiting for a spaceflight, my dad found the route to redemption on the back of an aging cartoon character.

From the afternoon in December 1966 that he first tried out the Manned Maneuvering Unit in a Martin Marietta simulator, he was hooked on a vision of a gas-propelled jetpack that would allow astronauts to operate outside their spacecraft. This vision had an obvious pop-culture antecedent. In the 1920s a comic-strip character named Buck Rogers—a rock-jawed, All-American World War I veteran—succumbed to the effects of a mysterious gas he encountered while working as a mine inspector. He fell into a deep sleep and woke after five centuries of slumber to a strange new world of spaceships, ray guns, and Asian over-lords. Though he initially traveled this new world via an antigravity belt, a device that allowed him and his best gal, Wilma, to leap great distances at a time, Buck eventually acquired a svelte and evidently omnidirec-tional jetpack. He eventually ventured into space in an adventure called *Tiger Men from Mars*, and his exploits in the cosmos changed America's vision of the future forever. Millions followed Buck's adventures in the funnies, on radio, and in movie serials. Among Buck's imitators and spiritual heirs are Flash Gordon, Brick Bradford, John Carter of Mars, and Han Solo.

A host of talented men and women spent significant amounts of time and money to wrestle that jetpack out of the funny papers and into the space shuttle. None worked harder, though, than Bruce McCandless and his chief collaborator, an Auburn-educated engineer and Air Force officer named Charles Edward ("Ed") Whitsett, Jr. Whitsett was a pale, bespectacled individual, mild-mannered but tenacious. He had a head start on my father. He'd been thinking and writing about jetpack technology as early as 1962. In a sense, he was trying to solve a problem that didn't exist yet: Namely, how could an astronaut venture outside his or her spaceship and perform constructive tasks in an environment with no oxygen, with extreme temperature fluctuations, and in an orbital "free fall" that would leave the spacefarer lolling in the practical equivalent of zero gravity? Alexei Leonov of the Soviet Union and American Ed White had proven that extravehicular activity was possible, that men could survive outside of their space capsule, but basically all they'd done was float. How could a man move from one part of a spaceship to another, or from one spacecraft to another craft, or from a spacecraft to a satellite, in order to make inspections or repairs? None of these needs really existed in the early sixties, when the programs of both nations were still just trying to fire tin cans into low Earth orbit and predict, more or less, where they would come back down. But clearly the needs would eventually arise, and various methods were proposed to address them.

In the mid-sixties, the Air Force assigned Whitsett to NASA to supervise development of the Air Force's Astronaut Maneuvering Unit. Gene Cernan's failed test flight of the AMU on Gemini 9 in 1966—the "spacewalk from hell," as Cernan called it—set the jetpack project back, but it never went away. McCandless, Whitsett, and a NASA engineer named Dave Schultz worked quietly but assiduously to keep the dream alive. They enlarged and improved the AMU all through the latter half of the decade and into the seventies. In the "Forgotten Astronauts" wire story that portrayed him as a washout in 1973, my dad mentioned the reason why he wanted to stay in the manned space program despite not having won a crew assignment on either Apollo or Skylab. "McCandless," said the article, "has helped develop the M509 experimental maneuvering unit. The Skylab astronauts strap it on like a backpack and propel themselves Buck Rogers–like around the Skylab interior. [He] wants to build a larger operational unit to perform space chores outside the shuttle." And that's exactly what he did.

Though the Skylab M509 tests in 1973 and 1974 were a resounding success, resulting in the triumph of the jetpack concept over both rocket boots and the handheld maneuvering unit, Whitsett and McCandless didn't rest on their laurels. Over the next several years, using whatever time and funding they could scrape together, the team made multiple upgrades—eleven, by one count—to what was now being called the "manned maneuvering unit," or MMU. The bulbous nitrogen-gas fuel tank of the ASMU was replaced with two streamlined aluminum tanks in the rear of the unit, each of which was wrapped in Kevlar. The number of propulsion nozzles was increased from fourteen to twenty-four, positioned around the jetpack to allow for six-degrees-of-freedom precision maneuvering. Smaller gyroscopes replaced those used on the ASMU, and, as space historian Andrew Chaikin has noted, the ASMU's "pistol-grip hand controllers, which were tiring to operate in pressurized spacesuit gloves, were replaced by small T-handles that needed just a nudge of the fingertips." The MMU's new arm units were made to be adjustable, to accommodate astronauts of all sizes. Painted white for maximum reflectivity, the unit was built to survive the 500-degree fluctuation in temperatures (from a high of 250 degrees F to a low of minus 250 F!) that an astronaut might encounter in space.

By 1980 the machine weighed in at 326 pounds. Like the AMU and the ASMU before it, the MMU was designed to fit with or "over" the astronaut's pressure suit. Shuttle astronauts wore a newly designed suit called the Extravehicular Maneuvering Unit, or EMU, a two-piece marvel of textile engineering made up of fourteen layers of Nylon ripstop, Gore-Tex, Kevlar, Mylar, and other substances. Power for the jetpack's electronics was supplied by two 16.8-volt silver-zinc batteries. Two motion-control handles—the translational hand controller and the rotational hand controller—were mounted on the unit's left and right armrests, respectively, and a button activated an "attitude-hold mode," which used motion-sensing gyroscopes to direct the firing of the thrusters to maintain an astronaut's position in space.

The machine had been tested in every way its designers could imagine. A representative of a local gun club visited Martin Marietta and shot the MMU's nitrogen fuel tank with a .50 caliber bullet to ascertain whether the tank would explode if pierced. (It didn't.) The jetpack was run through hundreds of hours of simulations. At my father's urging, a gifted and intense Martin Marietta project manager named Bill Bollendonk subjected the device to space-like conditions in the company's thermal vacuum facility. The MMU was no longer a "far out" experiment, as Mike Collins once called it. It was now a promising space tool. Unfortunately, for the moment, it was still an *unused* space tool. American astronauts remained on Earth, as NASA struggled to produce its next-generation orbital workhorse, the space shuttle.

As **with much of** both the American and Soviet space programs, the antecedents to the space shuttle can be found in Nazi Germany. The Nazis wanted a way to bomb America as a means to punish the United States for involving itself in World War II. But how to get a bomber all the way across the Atlantic and then back again? The answer, two German scientists said, was not to come back but to keep going—west, that is, where the bomber could land in a portion of the Pacific Ocean controlled by the Empire of Japan. These scientists envisioned a vehicle called the *Silbervogel*, or Silver Bird, a winged suborbital spacecraft that could be boosted out of Earth's atmosphere by rocket, then glide back down on a pair of stubby wings. As the atmosphere thickened, the gliding vehicle would be boosted, or bounce, back up again. In theory, this glide-and-hop sequence could be repeated several times and for several thousand miles, more than enough distance to allow for the fiery destruction of Times Square and eventual recovery of the spacecraft and its crew in the warm waters of the western ocean.

The *Silbervogel* never flew. The technical challenges involved in the project were insurmountable at the time. But the concept lived on: a space plane that could fly or be boosted by rocket beyond the atmosphere, return to

Earth under its own power, and be recovered and reused an indefinite number of times. This was the theory that animated the U.S. Air Force's X-20 Dyna-Soar (Dynamic Soarer) project, which lived from 1957 to 1963. The obvious difference of the Dyna-Soar from other space-vehicle designs at the time was its ability to glide back to Earth, as opposed to simply falling, as with the Mercury or Vostok capsules eventually used by the American and Soviet space efforts. Ultimately, though, the Dyna-Soar never flew either. Though detailed plans for its construction—involving molybdenum, graphite, and a nickel-based superalloy called René 41—were prepared, and a group of pilots, including Neil Armstrong, was selected for testing the craft, funding for the space glider was canceled in 1963. The nation focused instead on President Kennedy's goal of a lunar landing.

By 1972 enthusiasm and federal funds for manned space travel had dwindled dramatically. In announcing his vision for NASA's future, President Richard Nixon proposed a truncated space program—no missions to Mars or ambitious space stations—that would focus on developing a space plane similar in concept to the Dyna-Soar. Construction of shuttle-orbiter prototypes commenced in the mid-seventies. The first, named *Enterprise*, was produced in 1976 and used for reentry and landing tests. While early plans envisioned the orbiter being carried aloft by a hypersonic piloted aircraft, which could itself be reused any number of times, budget constraints led to development of a more mundane alternative: launch by rocket, with reusable boosters.

Despite such compromises, the shuttle, known to NASA as the Space Transportation System, held considerable promise. It was going to be a safe, economical, reusable space jet that would deploy and service satellites and, one day, assist in construction of a permanent space station. Talk of its capabilities and progress was more or less constant at our dinner table in the seventies. As familiar as the outline of the shuttle orbiter looks today, in the seventies the space plane seemed impossibly futuristic and sleek. At one point early in its development, Dad asked us to help him come up with names for the project. Like NASA, he wanted to express its functional, possibly commercial, applications. Among the possibilities the family suggested were the Earth Orbit Express, the Cosmic Carryall, and the Space Shipper. Fortunately, all of our suggestions were ignored.

My father was an avid onlooker and sometime participant in the shuttle's development. It was, in its own way, an engineering marvel. He took me out to JSC one afternoon to look at a prototype of the ceramic tiles that would be applied to the underbelly of the orbiter to protect it from the heat of reentry. He then asked a technician to heat the tile. An acetylene torch was produced, and the tile was heated till it glowed red.

"Go ahead," said Dad, grinning. "Pick it up."

I did. Amazingly, it was cool to the touch.

He loved stuff like that.

IN MY FATHER'S MIND, the space shuttle and his jetpack project were inextricably linked. The shuttle would get human beings into orbit, and the jetpack would allow them to leave the spacecraft and perform construction, maintenance, and repair work while they were there. Neither one made sense without the other. So Bruce McCandless stuck around. He worked hard, even on trivial assignments. He made himself a team player, as best he could, and obeyed all posted speed limits. The gigantic new astronaut class of 1978, the so-called Thirty-Five New Guys, must have eyed him suspiciously, wondering what albatross this oldster had killed to condemn him to a life wandering the halls of Building 4, haunted by the ghosts of canceled Apollo missions. But Bruce McCandless never complained about his lot, and as the old regime aged, new talent joined NASA and moved up in the hierarchy. Al Shepard and Deke Slayton moved on. George Abbey and John Young moved up. Abbey, a former Air Force officer and longtime NASA hand who served as director of JSC from 1996 to 2001, was a gifted but secretive administrator who became responsible, as director of flight operations, for choosing space shuttle crews. He attracted criticism for his management style, which some called insular and arbitrary. Whatever his flaws, Abbey's ascendance at NASA changed the metrics involved in getting assigned to a spaceflight. It didn't matter as much anymore if you'd been a test pilot, or gone bear hunting with Deke, or could hit 105 mph on the Gulf Freeway in your gold Corvette. As the hardware and software required for development of new spacecraft and satellite technologies evolved, engineering prowess became increasingly valuable.

Bruce McCandless succeeded by turning conference-room concepts into things that actually worked in space. He knew the theories behind celestial mechanics, but he was also adept with a socket wrench and a power drill. According to fellow astronaut Kathy Sullivan, McCandless was a "brilliant engineer, with a photographic memory and an extraordinary capacity for detail." When considering new hardware NASA might need for future missions, he knew "exactly which type and size of fastener or part the new gadget called for, what metals or synthetic materials were best suited for the purpose (along with what physical properties made them so), and which vendors, by name, would be good sources for each component." My father, in other words, turned himself into an expert on the tools and techniques needed for space repairs. So far, so good. He was becoming a valuable commodity. Still, more than a decade into his NASA career, what he really wanted was a chance to put his expertise to use—and there was no sign of that happening anytime soon.

THE MMU GOT A big boost in 1979, as NASA was working to bring the shuttle online. When the orbiter *Columbia* was transported on the back of a specially modified 747 from its California birthplace to Florida for eventual launch, an alarming number of thermal insulation tiles disappeared from the spacecraft during the flight. As the tiles were necessary to protect the orbiter from temperatures as high as 3,000 degrees Fahrenheit during reentry into the Earth's atmosphere, NASA officials realized they had a problem on their hands. Even a single missing tile in the wrong place—on the orbiter's underside, for example, which it used as a heat shield on reentry, or on the leading edges of the shuttle's wings—could constitute a disastrous chink in the machine's thermal armor. As one NASA administrator put it, it was time for the agency to do some "soul searching" about the problem.

One possible solution was outfitting an astronaut with a manned maneuvering unit and a sort of thermal tile patch kit for in-orbit repair procedures. On October 2, 1979, NASA announced that it would "proceed with an accelerated development of the Manned Maneuvering Unit that will allow an astronaut to inspect and repair the Space Shuttle's heat resistant tiles while in orbit." After having spent several years in an extended design phase— what my dad later called the "lean years"—the MMU suddenly became a hot property. McCandless, Whitsett, and Martin Marietta hastily finalized a proposal and specifications for their prized machine. NASA awarded a $26 million MMU production contract to Martin Marietta in 1980, the same year it awarded the company a $2 million contract for development of an on-orbit thermal tile repair kit. Astronauts John Young and Bob Crippen, who were scheduled to fly the first shuttle mission, participated in simulations of a tile-repair procedure using the manned maneuvering unit at the Martin Marietta facility in Denver. In the meantime, though, the orbiter was completely retiled, and engineers were hopeful that missing tiles would no longer be a problem. Thus, though the thermal-tile emergency contributed to development of the jetpack, the MMU as delivered lacked a purpose.

This is where Solar Max came in.

On Christmas Day 1980, Bruce McCandless wrote to his brother Douglas that "the Shuttle is in fact nearing readiness for its first launch." He also reported that production of two MMU prototypes was coming along and that he expected to start qualification testing of the units soon. Despite the fact that the major impetus to the MMU's production had apparently gone away, Bruce McCandless refused to give up. NASA was a different place in 1979 than it had been ten years earlier. In the wake of congressional abandonment of space exploration as a national priority halfway through the Apollo program, the agency began to push its undertakings not as wonders of science and curiosity in their own right but as economic projects designed and built for their utility. The space shuttle was peddled as a cosmic-step van, meant to allow for transport, deployment, and repair of satellites and space habitations. Everything was seen through the prism of practicality. Funding a project was no longer just a matter of demonstrating what it could do but showing why it should be done in the first place.

In the case of the MMU, Bruce McCandless now linked the *why* to a balky little satellite called Solar Max. NASA launched the Solar Maximum Mission in 1980. Intended to study atmospheric phenomena on the sun, particularly solar flares, the satellite functioned well for a period of six months, then effectively died as a result of a number of electrical and attitude-control malfunctions. By the end of 1980 it appeared that the United States had spent millions of dollars on a piece of space hardware that had worked for approximately the length of a Major League Baseball season. Now it was spinning around the Earth in useless orbit, an embarrassment to NASA and Congress and a frustration to the scientists who were counting on data from the device.

Bruce McCandless hitched his jetpack to this faulty artificial eye. The manned maneuvering unit, he explained, would allow astronauts to exit the space shuttle, proceed under their own power to the satellite, and make the fixes necessary to get Solar Max functioning again. It was a simple plan but a good one. There was just one problem. No one knew yet whether the shuttle *itself* would work. Fortunately, that question was about to be answered.

THE FIRST FULL-FLEDGED FLIGHT of the shuttle occurred on April 12, 1981. The orbiter *Columbia* lifted off from Kennedy Space Center, piloted by the gifted but taciturn moonwalker John Young and his rookie partner, native Texan Bob Crippen. It spent a little more than two days in space, orbited the Earth thirty-six times, and then prepared for its baptism by fire, reentry into Earth's atmosphere and a landing at Edwards Air Force Base in California. The mission was a nail-biter to those involved. It was the first time in NASA's history that the functionality of a spacecraft would be demonstrated by means of a crewed orbital flight—in essence a test of a huge, extremely complicated machine, only with two human lives on the line. John Young's wife, Susy, didn't expect him to survive. Young himself estimated his chance of death at 50 percent. Given the risks in life, investment, and prestige involved, it wasn't surprising that a sizable portion of the world's population tuned in on television to watch, expecting disaster, as Young and Crippen brought *Columbia* home.

I witnessed the proceedings that day. It wasn't originally my plan. I was sitting in a Marxist literature class at the University of Texas when my professor, a genial cynic named Jim Schmitt, mentioned the mission. He noted, incredulously, that one of his colleagues had suggested attendance at class might be sparse that afternoon in light of the scheduled landing.

"You all aren't interested in that nonsense, are you?" he asked. The room was silent. "Whitey's on the Moon and all that? Good Lord. Maybe you are. How many of you want to see the space shuttle land?" Almost every hand shot up. Shaking his head, the professor bade us young Marxists farewell, and we piled out into the sunshine of an April afternoon in search of television sets to witness what we hoped (or feared, given the makeup of the class) would be the triumph of capitalist America's most visible symbol of technological wizardry. The landing went perfectly. In fact, it was twenty-seven seconds ahead of schedule. Loudspeakers broadcast the radio transmissions between the astronauts and mission control to the 250,000 spectators on hand. Flags flew. Music played. "New Era Ushered In By Shuttle," wrote the *Chicago Tribune*. NASA was back.

The space shuttle never really became the dream ship it was designed to be. It was far more expensive to produce and maintain than initially envisioned, and it was plagued by safety problems that eventually killed fourteen astronauts on two missions. Still, the thirty-year shuttle program was an important and largely successful stage of American space exploration. The five shuttle orbiters—*Columbia, Challenger, Discovery, Endeavour,* and *Atlantis*—flew 135 missions from 1981 through 2011, covering 542 million miles and 21,152 orbits of Earth. The vessels deployed 180 satellites, helped to construct the International Space Station, and docked nine times with the Soviet (later Russian) space station, *Mir.* Though never quite as startling in its design or bold in its execution as the hardware of Apollo, the shuttle had its own brand of beauty. It was, as astronaut Story Musgrave put it, a butterfly bolted to a bullet.

ONCE THE SHUTTLE PROVED flight-worthy, Bruce McCandless redoubled his efforts to get a seat on board. He kept a journal of the campaign. Characteristically, there is not a word of recrimination or frustration in it, though he must have been tempted to vent on numerous occasions. Much of the text is technical—notes and numbers scrawled in my dad's tiny but insistent print, as incomprehensible to the layman as an alchemist's gnomic inscriptions. But there are practical and even political notes as well. In May 1981 he wrote that in his opinion, a successful servicing of the Solar Max would "arouse same order of public interest as going to moon." On May 22 he met with Gene Kranz, one of the heroes of Apollo 13, a thick-handed man whose features appeared to have been chipped out of a cinder block. As deputy director of mission operations, Kranz effectively controlled who got to do what on every mission. Kranz, my father wrote, "liked the idea of 2 MMUs—rescue, etc.," but was "concerned about MMU phase plane logic for stabilization." On June 2, 1981, just two months after the successful maiden flight of the shuttle, McCandless recorded a piece of good news. One of the flight directors who was skeptical of the MMU's potential had gone to Chris Kraft to torpedo the project. "Kraft said go do it," Dad wrote. "This is a fantastic mission. [Flight Director] went to Kraft—[Kraft] chewed his ass—said GO!"

On June 29 he noted intelligence he received regarding a meeting attended by NASA deputy administrator Hans Mark. Mark, said the source, indicated that NASA was "in desperate straits to demonstrate utility of shuttle in early flights ... Don Turner said looks like ready for brief on Solar Max Mission. Yes." And even as he was lobbying hard for a Solar Max repair mission, Dad was working on improvements to the MMU. His journal is filled with cost estimates, technical specs, secret reports, and meeting notes. "Test flight [of the MMU]," he reminds himself, "is an essential part of Solar Max Repair Mission & should not be split out."

By early 1982, having won NASA support for the project, McCandless and Whitsett were concerned about Congress. Representative Ronnie Flippo of Alabama, a prominent member of the House's Science and Technology Committee, had "zeroed us out" of the budget, he wrote, and the project needed "Astro advocates" like Apollo 17 astronaut, now New Mexico senator, Jack Schmitt to help out with the funding issue. Flippo was a backer of a project called the Solar Polar mission and apparently believed that the Solar Max repair effort was taking money (an estimated $57 million launch cost) away from his project. Schmitt was initially ambivalent about the Solar Max rescue mission but ultimately helpful. My father's friend Jack Lousma, fresh off his successful command of the third shuttle mission, "asked Schmitt privately—without any sales pitch or bullshit—is Solar Max going to get approval? Ans. Yes—Primary objection was the way it was presented to the Congress as a *demonstration* . . . NASA must tell the Congress it wants to go fix a scientific satellite that failed . . . and it will get approval." [Emphasis added.] On July 9, 1982, Dad met with Gene Kranz again. Kranz was not only supportive. He now said that it might be possible to move the demonstration flight up to STS-8. He also was "very interested in having heart rate available in control center—useful in giving 'GO' for untether & for instructing MMU pilot to 'be able to tell him to slow down or rest awhile.'"

NASA's brass seemed increasingly comfortable with the MMU. In truth, the machine was hard to ignore. It was a combination of technical wizardry and imagination-grabbing potential that the agency needed as it vied for a place in the public's attention. In keeping with its new, stripped-down, cost-sensitive ethos, NASA downplayed the romantic aspects of space travel when describing the shuttle, emphasizing its prosaic qualities instead. In a similar vein, the MMU was characterized by the agency in utilitarian terms, as a tool for inspecting and servicing satellites. Whitsett defined the MMU as "a self-contained propulsive backpack" and "a miniature space-craft which an astronaut straps on for spacewalking." Shuttle astronaut Joe Allen called it "[the] spaceship's special dinghy," which "resembles a back-pack with armrests, or some kind of overstuffed rocket chair." In a brochure for the "payload community," that is, third parties willing to pay for cargo space in the shuttle, NASA advertised the MMU like this:

> Since the Manned Maneuvering Unit has a six-degree-of-freedom control authority, an automatic attitude-hold capability and electrical outlets for such ancillary equipment as power tools, a portable light, cameras and instrument monitoring devices, the unit is quite versatile and adaptable to many payload task requirements.

Less impressed, as they usually are, French journalists dubbed the device the "flying armchair."

After many years of design, simulation, advocacy, and waiting, the MMU was ready. What was less clear was whether its co-designer would be allowed to fly it. The astronaut corps was bigger than ever and as talented as always. My father tried not to think about his disappointments during the Apollo era and focused instead on making the machine the best it could be. In the meantime, the sphinxlike George Abbey was wrestling with shuttle mission crew assignments. It wasn't until February 1983 that the official announcement came. It was scripted in NASA's bland language of bureaucracy, as if the agency were detailing its copier maintenance contracts. The crew of Mission STS-11 would test an intriguing and highly practical device called the Manned Maneuvering Unit. Primary pilot of the machine, said NASA, would be longtime agency veteran Bruce McCandless.

I WON'T SAY FINALLY getting a flight assignment made Bruce McCandless a happy man, full stop. He was always too driven to be entirely satisfied. What I will say is that it changed him. In May 1983 he and my mom drove to Austin to see me graduate from the University of Texas. On our way to the commencement ceremony, late, as usual, I realized I'd left my graduation cap—the "mortarboard"—back in my apartment. I spoke up sheepishly from the backseat. Dad never hesitated. We were headed south on the Drag, Austin's central campus thoroughfare, and not an easy place to drive. In the middle of traffic he executed a screaming 180-degree turn that would have made Bobby Unser proud, stomped on the gas, and headed north toward my apartment. The squeal of rubber on the hot pavement was like a scream of release. Motorists honked in protest. A unicyclist in the crosswalk saw his life flash before his eyes. I couldn't help laughing. Remarkably, my father laughed too. Our campus escape velocity that day probably didn't top fifty, but it was good to see the old Grim Reaper emerge from his sports jacketed-shell. We retrieved the mortarboard and made it to the ceremony on time.

Despite NASA's emotionless description of the mission, the media soon grasped what was going to happen. The stories started coming as the year wore on and the shuttle made several successful flights. By December, the public was interested.

ASTRONAUT TO FACE SPACE ALONE, trumpeted the *Houston Chronicle*. Buck Rogers was about to fly.

14

THE ACCIDENTAL ICON

In order to talk about Bruce McCandless's first spaceflight, I have to backtrack a bit. On Christmas Eve 1979, elements of the Soviet Union's 40th Army rumbled into Afghanistan, heading for the capital city of Kabul. The Kremlin commenced the invasion to prop up Afghanistan's communist regime, which faced armed resistance from the country's devoutly Muslim rural population. The Soviet Union claimed it was acting in accordance with the so-called Brezhnev Doctrine, which classified an attack on any socialist government aligned with Moscow as an attack on all such governments, and sanctioned military action in response. Western nations condemned the incursion. Some stopped at words, but the United States sent money and arms to equip the Islamic rebels, who called themselves the *mujahideen*. Over the course of the next several years, millions of Afghan civilians died or were displaced as the Soviet army and its special forces fought the *mujahideen* for control of the rugged countryside. It was Vietnam all over again, only this time it was Washington's turn to support an oppressed native people in their fight against a technologically superior imperialist enemy.

The Cold War had resumed.

Unlike the most *intense* years of the Cold War, though—the period of Sputnik, the Berlin Wall, and the Cuban Missile Crisis—the space race was mostly an afterthought at this point. The Soviet space program was formidable and had notched a number of impressive firsts, including not only sending the first man and woman into space but also more technically challenging achievements like sending back the first images from Venus, in 1975. Still, the Russian effort never quite recovered from its failure to develop a workhorse rocket capable, like von Braun's massive Saturn V, of sending a man to the moon. Tests of the N-1, Moscow's response to von Braun, repeatedly failed. It didn't help the Soviet image that the nation's attempt to develop a space shuttle of its own ended up looking exactly like NASA's version—an unambiguous, if unspoken, confession that Soviet designers preferred to borrow American engineering rather than trust their own. The Soviet shuttle, the *Buran*, or Blizzard, was eventually trotted out for public viewing at the 1989 Paris Air Show. It even made a successful wholly automated flight, which the space shuttle never did. All in all, though, *Buran* was an expensive dead end for the Soviets. The orbiter died an ignominious death, buried in the collapse of a hangar during a particularly vicious storm at the Baikonur Cosmodrome in 2002. The Soviets, and later the Russians, set their sights instead on more modest goals and stuck with flying the less sophisticated but more reliable *Soyuz* spacecraft, which looks on landing a bit like a giant black champagne cork.

America and the Soviet Union played power politics all through the eighties. Under President Reagan, the United States not only armed Afghan rebels but also funded anticommunist fighters in Nicaragua and mounted a political and public relations offensive in Europe for approval of installation of cruise missiles on the continent. And since America had an edge in aerospace technology, Reagan took advantage of the opportunity to militarize and, possibly, colonize space. In 1983 he called for development of the Strategic Defense Initiative, sometimes known as the "Star Wars" defense system, which proposed to use lasers and particle-beam weapons to destroy intercontinental ballistic missiles launched at the United States. In his January 1984 State of the Union address, the president announced plans for construction of the Freedom Space Station, an ambitious American response to the Soviets' periodically manned Salyut orbital outposts. Freedom went through a number of planning stages. Like the Space Transportation System, it lost pieces as it proceeded, growing smaller and more modest over time. It was eventually absorbed into plans for the International

Space Station (ISS), the first module of which was deployed some fourteen years after Reagan's speech, in 1998. Ironically, the ISS became a symbol of American-Russian cooperation.

IT WAS FITTING IN this decade of increased militarism that Bruce McCandless's first space-flight boasted a bold and photogenic crew—the Right Stuff, with sideburns. Commander Vance Brand was fifty-three, an easygoing former Marine test pilot. A Group 5 astronaut like my father, Brand had previously flown on the Apollo-Soyuz mission in 1975 and, in fact, barely survived the trip. He inhaled hydrazine and nitrogen tetroxide fumes upon reentry and was hospitalized for three weeks to recover. He'd also commanded the fifth space shuttle mission. Pilot Robert L. "Hoot" Gibson was a Naval Academy graduate and one of the best fliers NASA ever employed, a man who, in the words of one of his colleagues, *wore* his airplane rather than occupied it. An improbably handsome individual with a lush mustache, Gibson lent the group a swashbuckling flair.

There were three mission specialists. Lt. Colonel Bob Stewart was scheduled to test the MMU along with my father. The first active-duty Army officer to go into space, Stewart was a soft-spoken, cerebral combat veteran who won the Distinguished Flying Cross for service in Vietnam. He later became a test pilot and helped to develop the Army's innovative and agile Black Hawk tactical transport helicopter. Dr. Ron McNair, a civilian, was a brainy but affable MIT-trained astrophysicist from South Carolina who, as a child, was threatened with arrest for trying to check out a book from a local library. Just thirty-three years old, he held a black belt in karate and was an accomplished jazz saxophonist to boot—exactly the kind of gifted eccentric my dad admired and wanted to be. Finally, there was Bruce McCandless, one of the comparatively few men in the world who could claim to have jumped from a burning airplane into the ocean and lived to tell about it. Forty-six years old, his glossy dark hair gone wispy and white, the slender kid genius who'd joined the astronaut corps in 1966 was now described by one journalist as "almost grandfatherly." It was a close-knit crew, Gibson later recalled. As it turned out, it would need to be.

The five men rose early on the morning of February 3, 1984, and joined George Abbey and John Young for breakfast. Afterward they received a weather briefing, donned their flight suits, and underwent a prelaunch medical examination. After the traditional walkout to greet assembled onlookers, the men boarded the Astrovan for the nine-mile drive to Kennedy Space Center's Launchpad 39A. Here's where things got real. The orbiter stood in the distance as if impatient to leave, brilliant in the illumination of the launch site's xenon spotlights. The closer the van approached, the bigger the 200-foot spacecraft seemed. It loomed above them like some giant altar, and, at this point, when there was no real chance to turn back, the astronauts wondered whether they would turn out to be the officiants or an offering. Bruce McCandless's heartbeat quickened. He told himself to take measured breaths, remember his training, focus on the task at hand. The world suddenly seemed more vivid than he recalled, a place of sharp edges and primary colors. The driver said something he didn't quite catch. Out of the van now and up the elevator the astronauts went, the rattle and clank and nervous energy of launch prep all around them, loudspeakers blaring checklist instructions, water vapor swirling around the liquid oxygen vented from the main engine nozzles, the shuttle stack hissing like a living thing, like a racehorse restless and snorting in its stall. Finally the men climbed into the orbiter and found their seats. Brand and Pilot Hoot Gibson cycled through their launch preparations. The mission specialists listened to chatter on the Launch Control Center frequency and waited.

STS-41B left Earth at seven that morning, the orbiter shuddering skyward atop a barely controlled cataclysm: the seven million pounds of white-hot thrust generated by the orbiter's three main engines and two solid-fuel rocket boosters. Almost immediately after takeoff, the shuttle rolled to head east, taking advantage of Earth's rotation to add relative speed to its trajectory. My father occupied Seat 5, alone on the orbiter's mid-deck, secured in place by lap belts and shoulder straps. "I was in my seat," he remembered, "locked back, looking upward. The initial thrust produced only about 1.5 gs of acceleration. As the solid-fuel rockets were consumed and fell away and our fuel was consumed, the acceleration level built up until it reached three times the force of gravity. It felt as if a big hand was in the middle of my back, giving me a mighty push." Though the g force he experienced wasn't as strong as that produced by one of the old Saturn V rockets, the vibrations of launch and ascent were nevertheless intense. Kathy Sullivan once described the sensation as akin to being embedded in an earthquake. Dad likened the shaking and jolts to a train wreck. It felt like *Challenger* was going to break apart, and he shut his mouth tight so his stomach wouldn't fall out. At around thirty-five seconds after liftoff, the main engines throttled back as the shuttle approached Mach 1 and punched through the sound barrier. The din lessened as the air grew thinner, and soon the main engines powered up again, pushing the shuttle to "Max Q," or maximum aerodynamic pressure on the spacecraft, at around a minute after liftoff, roughly seven miles up. As the atmospheric pressure eased, the spacecraft's velocity steadily increased. The astronauts felt the change distinctly, as the solid rocket booster thrust level also increased due to the propellant's grain design. The orbiter was moving at 3,000 miles per hour.

Two minutes into ascent, the spent solid rocket boosters were blasted away to parachute back down into the Atlantic Ocean. Though the orbiter continued to accelerate, moving past 8,000—no, now 10,000—miles per hour, the ride grew smoother as the spacecraft hauled itself up out of Earth's soupy atmosphere. At fifty miles above the planet, Bruce McCandless finally reached space, by NASA standards, and earned his gold astronaut pin. There was no one with him to record it, but I like to think that a smile spread across his face. He didn't have much time to think about the accomplishment, though. The next milestone was MECO, or Main Engine Cut-Off. By this point the spacecraft had attained sufficient altitude and velocity—approximately 17,500 mph—to reach orbit without aid from its three main engines. The engines were shut down, and the orbiter jettisoned its rust-colored external fuel tank. Another nudge of thrust, this time from the shuttle's relatively small orbital maneuvering system thrusters, placed *Challenger* in its planned orbit. Anything not secured was now floating, liberated by the weightlessness of free fall. The entire ascent had taken only a few minutes. Only Vance Brand had experienced the feeling of weight-lessness before, so the crew spent some time orienting themselves. They grinned at each other's awkwardness and glee. They unpacked and tested their gear and the orbiter's navigational and life-support systems. And then they got to work.

Unfortunately, very little proceeded as planned. The mission's first assignment was to deploy two commercial communications satellites, the Indonesian Palapa B-2 and Western Union's Westar VI. The deployments were money-making enterprises in themselves, of course, but they were also meant to prove that the shuttle was a relatively cheap and reliable way to get hardware into space. The orbital "launches" were planned for consecutive days. In each case, the communication satellite left *Challenger* in nominal fashion. Everything checked out fine. But the satellites' booster rockets failed to function as they were supposed to, with the result that the two satellites ended up stuck in orbits much lower than the 22,300 miles above Earth they were intended for. Bottom line: they were useless. NASA hadn't built the satellites. Their failure wasn't the agency's fault, and it certainly wasn't the fault of the crew, but it got the mission off to a troubling start nevertheless.

The bad luck continued. A test of shuttle rendezvous techniques was aborted when the Integrated Rendezvous Target, a Mylar balloon bal-lasted with lead, meant to be released and thereafter serve as a target for the orbiter's tracking and navigation systems, exploded instead of inflat-ing properly. Later on in the flight, the mechanical "wrist" on the orbiter's Canadian-made remote manipulator arm malfunctioned, which prevented a scheduled test of a number of satellite capture techniques. The onboard toilet broke. A buildup of ice on *Challenger*'s port side broke off during reen-try and barely missed smashing into the vessel's orbital maneuvering system engine pod. Such a collision would almost certainly have exposed the fuel tanks to something like 3,000 degrees Fahrenheit of heat, and the resultant explosion would have destroyed the spacecraft. It was a significant hazard, but completely unknown to the crew until after they'd landed. Damage to the orbiter's heat-resistant tiles was also observed after the shuttle returned to Earth, and *Challenger*'s brakes partially failed at touchdown. Indeed, the Soviet news agency TASS soberly informed its readers that STS-41B was the least successful mission of the American space shuttle program thus far.

Such was the setting for the first test of the MMU. In the face of faulty rocket engines, exploding balloons, and poor publicity, NASA desperately needed a win. On February 7 Bruce McCandless and Bob Stewart woke and ate a quick breakfast. They spent a little over an hour prebreathing oxygen to purge the nitrogen from their systems, and then sent biomedical readings—temperature, heart rate, and so on—down to Houston for review. Once the readings were approved by Mission Control, the astronauts worked themselves into and powered up their EMUs, or pressure suits. They entered *Challenger*'s airlock and moved through the structural and thermal hatches into the orbiter's payload bay. The two men used safety tethers at this point, which was all, besides their own hands, that kept them from drifting off into the apparently limitless universe around them. McCandless's EMU had a horizontal red stripe just above each knee, while Stewart's was unadorned.

There were two MMUs on *Challenger*. Martin Marietta had actually built three machines, but the first, Unit 1, was meant for testing and qualification purposes only, and therefore never left Earth. The two operational jetpacks were designated as Units 2 and 3, with one mounted on each side of the orbiter's payload bay. McCandless and Stewart checked out the nitrogen gas pressure and battery-power levels on the machines and reported their findings to Houston. *Challenger* was flying with its tail toward Earth, forty degrees off vertical. Right around the time the shuttle passed over Namibia, in Africa, Bruce McCandless backed into Unit 3 and fastened himself in. Upon receiving the all-clear from Vance Brand, he launched himself by leaning forward and lowering his toes. That was all it took. Untethered, he began to move away from the payload bay wall.

The words came shortly afterward: "That's one small step for Neil, but it's a heckuva leap for me."

He was suffering from symptoms of what was either a head cold or space adaptation syndrome during the early stages of the mission, but he wasn't prepared to admit it to Mission Control for fear that Houston would scrub his test flight. Bruce McCandless had waited forty-six years to get to space, and he wasn't going to let the moment go. He started down his flight-test checklist by confirming the unit's maneuvering capabilities: "pitch down, pitch up, roll right, roll left, translate forward, translate aft— the entire sequence designed to exercise all axes." Everything worked. The MMU was ready to fly. As he started moving away from *Challenger*, doing a leisurely waltz with eternity, McCandless took in his surroundings. He tried to stay calm, but his voice gave him away. "I'm trying to figure out what sort of landmass we're coming up on," he commented. "Looks like Florida. It IS Florida! It's the Cape!"

Commander Vance Brand observed that my father was now the "world's fastest man," as he was traveling at a speed relative to Earth of some five miles per second (although, of course, so was Brand). McCandless moved out 150 feet from the shuttle and returned, then traveled out to roughly 300 feet, and came back. I say *roughly* because the device he used to measure the distance was a slender bar of aluminum—sometimes referred to as the "solid-state range finder" or "state-of-the-art range finder"—that had marks scored on it representing the relative size of the shuttle payload bay when viewed from 100, 200, and 300 feet. Several years later, I was with my father at a dinner party in Nassau Bay attended by Alexei Leonov, the first man to walk in space, and a veteran of the Soviets' comparatively low-tech approach to orbital travel. My dad told him how the solid-state range finder worked, and the two exchanged a hearty laugh.

"Again stealing Russian technology!" said Leonov.

After his long-distance sorties, Bruce McCandless returned to the payload bay, docked the jetpack, and began recharging its supply of nitrogen propellant. Bob Stewart had been working in the meantime on testing the Manipulator Foot Restraint, or MFR, which was a platform to secure the astronauts' feet as they were maneuvered about by the remote manipulator arm. The Army officer was having trouble with the equipment. My father took over the MFR test, and Stewart switched with him to continue refueling the MMU. Once it was fully recharged, Stewart conducted his own MMU test flight, also proceeding out to 150 and then to 300 feet before returning to the orbiter.

While Stewart was flying, my dad worked his way into the MFR and practiced satellite capture and repair techniques as Ron McNair, inside the orbiter, operated the 55-foot-long robotic arm. It was like being the lineman in a utility company's cherry picker, Dad said. He worked some with the Trunion Pad Attachment Device, or TPAD, practicing docking with a Solar Max–like setup on the German-made pallet satellite called the SPAS, and then conducted a force-resistance activity to test the stability of the robotic arm. All in all, it was a productive day.

It was also a photogenic day. Back on Earth, viewers were captivated by the video images of McCandless and Stewart ghosting away into space. Capcom Jerry Ross reported the next morning that "we certainly had a good time watching you all yesterday, and you were the talk of the world this morning." McCandless Backpacks Into Space History, said one newspaper headline. And another: Free Flight Gives McCandless Spirit of Those Who Dare.

It was official. NASA had a win.

McCandless and Stewart went back out in the jetpacks on February 9, Day 6 of the mission. The goal this time was to practice the way future astronauts would capture, by latching onto, the Solar Max satellite. Dad started early, around 4:40 a.m. Houston time. He climbed into MMU Unit 2, disconnected his safety tether, and proceeded to execute a series of somersaults above the payload bay. More Solar Max capture simulations followed. Bob Stewart then donned and tested the jetpack. As his time in the machine was concluding, a portable foot restraint floated out of the payload bay. Stewart wasn't completely out of the MMU, so he offered to retrieve it. My dad and Vance Brand chose instead to maneuver the orbiter close enough to the object so my father could reach it. Bruce McCandless was tethered and holding on to one of the orbiter's starboard slide wires with one hand when he stretched out the other like a fan reaching for a foul ball and was able to grab the restraint. It was an unplanned but effective demonstration of the giant space vehicle's precise maneuverability—proof that *Challenger* could move like a mouse in a china shop—and the procedure earned a round of applause from Mission Control. Stewart moved on to conduct a test to determine the feasibility of refueling satellites, with Freon 22 standing in for the more volatile, and toxic, hydrazine. After my father conducted a few more engineering tests of the MMU, the EVA concluded. And just in time: President Reagan was on the line, calling to congratulate the crew.

REAGAN: Good morning to you [Commander Brand] and your crew. I'm talking to you from California. I don't know exactly where you are, I know you're up there someplace, but you're all doing a fine job on this historic mission, and I'd like to say hello to Bruce McCandless and Bob Stewart, for sending us this spectacular television coverage of man's historic walk in space. Let me ask you, what's it like to work out there, unattached to the shuttle, and maneuvering freely in space?

MCCANDLESS: Well we've had a great deal of training, sir, so it feels quite comfortable. The view is simply spectacular and panoramic and we believe every unit's first time working unattached we're literally opening a new frontier in what man can do in space, and we'll be paving the way for many important operations on the coming space station, sir.

```
REAGAN: Well that is just great. You've really opened
a new era for the world in space with this mission.
You've shown both our commercial partners and our for-
eign partners, who play an important role in this and
other missions to come, that man does have the tools
to work effectively in space.
```

This was part of the official exchange, and it's appropriately decorous. But the circumstances surrounding the in-space conversation with the White House that day are more entertaining. My dad had just finished his second EVA when word came that President Reagan wanted to speak to the crew via video link. Vance Brand promptly agreed to the request, if it *was* a request, and started planning how to stage the conversation. Brand, Hoot Gibson, and Ron McNair arranged themselves in the cabin against the windows, with a TV camera looking at and past them into the payload bay. Brand directed Bob Stewart to take up a position in the field of view holding on to the shuttle, and told my dad to "fly out a ways" and then stop, so the president could see an astronaut floating in space. A good astronaut never passes up a chance to sell the program, and this was a canny deployment of assets. But there was one small problem: The MMU was out of fuel. My father later recounted that he didn't want to disappoint Vance or the president, so he flew out about ten or fifteen feet and very gingerly nulled all of his rates so that he wasn't drifting in any direction. He then switched off both electrical systems, killing the MMU so that it couldn't fire any thrusters, and he hung in that mode while the president talked. Reagan announced the nation's commitment to building a permanently manned space station. Bruce McCandless replied politely, but all the while he was worried he would start drifting out of position, or, if Vance fired a thruster on the shuttle, the orbiter would definitely start moving away, or that he would have a little bit of imbalance and tilt over backward and "moon" the president. But none of that happened, and as soon as the president signed off, McCandless switched on the MMU control system, looked for the closest piece of the orbiter, pointed at it, put the hand controller in +X, got a

sort of sighing noise as it accelerated in that direction, and totally ran out of
propellant. Fortunately, the last sigh did the job. It drove him into contact
with the orbiter. He grabbed hold and moved hand-over-hand along a rail
until he got back to the donning station and shut down the MMU again.

This day, like the first, had been a smashing success, a dream come true
for Bruce McCandless. My father was a serious man, saturnine even, on
most occasions. In a celebratory mood, though, he could be downright
expansive. There was a fair amount of joking on the flight, much of which
involved the hazing given Army officer Stewart by Naval Academy grads
McCandless and Gibson. At one point toward the end of the EVA, it
became evident that the camera on Bruce McCandless's helmet was malfunctioning. He switched it on and off to no avail. He then offered to hit
himself in the head with a hammer to fix it. While Houston discouraged
this procedure, Stewart evidently was allowed to do the honors.

ASTRONAUTS WHO HAVE STEPPED into space have written about the
experience as the best part of their journey. Ed White, the first American to
perform an EVA, didn't want it to end. He said that reentering the capsule
was the saddest moment of his life. One can imagine. My father wasn't an
effusive man, but he felt the enormity of it all. He tried to focus, but the
prospect registered in his flesh: The elevated heartbeat. The sweating palms.
How could it not? It was February of 1984 when Bruce McCandless stepped
outside, and . . . no. That's not it. He didn't *step*. He didn't float, either, since he
was propelled, ever so slightly, by the child-like puffs of nitrogen gas from his
jetpack. Like a sloop leaving a harbor, he *sailed* beyond the payload bay and
there it was, the ultimate emptiness, eternal absence but nestled inside it like
a tiny pearl in an infinite shell Earth in every shade of blue and all of them
at once, cobalt clarion clear as if freshly made and all the colors sharper here
than the cleanest cold front day below. He sailed out. He reapproached. He
turned somersaults. He dipped left, dipped right, and eventually he returned

to *Challenger*. In through the airlock and back on board, he shed his Kevlar carapace and shrugged into his flight suit, relieved to bathe again in oxygen, elated, really, and part of him still outside but inwardly reassured to be among his crewmates, the warmth of their bodies, the jokes offered up like small gifts, the sounds of their shared existence. The mission continued. By day the deserts passed below their spaceship gold and burning brown and seas stippled with shadow and sediment and the great whispering forests of the continents pocked by steel-gray asphalt, but at night the cities wearing webs of light emerged as if from out of the ocean depths and shone like tiny constellations. Where one portion of the globe was clear, on another the astronauts watched bursts of electric white pulse along irregular pathways in silvered carpets of cloud. *Lightning*, some said, but my father knew better: these soft explosions were Earth's synapses firing, her slumbering children inventing the future as their minds ventured out from inside their heads and their dreams slipped through the waves of water vapor like dolphins playing in the surf. If life is, as some scientists posit, reducible to a few simple elements—a protective membrane enclosing a method of metabolism and the ability to reproduce—surely the planet itself is a living thing: the atmosphere its shimmering shield, photosynthesis its metabolism, ourselves the living seed, the lumpy pink and brown collections of protein codes it now disperses spore-like into the heavens in sleek metal cases. A *feeling*, this. Not a thought. Not for recording. And still Bruce McCandless watched, and he wondered, and eventually he consigned himself to sleep as well, floating as he fell on his arc across the sky, content to leave the organism below him to its longings and its great green visions and deep in the dark its hope and lonely promises.

THE JETPACK PERFORMED BEAUTIFULLY. "Pass on to Ed [Whitsett]," said my dad early in the flight on the first day, "we sure have a nice flying machine here." It wasn't completely like operating in the simulator. The machine "shuttered, rattled, and shaked" as it flew, which turned out to be because, as Andrew Chaikin put it, "The thrusters were designed to fire through the jetpack's center of gravity, which was offset slightly from that of the suited astronaut. As a result, each firing produced a small rotational force, which the MMU's sensitive attitude-control system had to counteract [by firing a thruster or thrusters in response]."

The biggest issue for Bruce McCandless on his initial EVA was that his pressure suit was hypercooled to avoid the problem Gene Cernan encountered on Gemini 9, when he overheated while battling weightlessness and his umbilical tether to get to and test the old Astronaut Maneuvering Unit. Because my dad was expending very little energy once he was in the MMU, the pressure suit grew chilly. In fact, his teeth were chattering as he moved farther away from the shuttle—a problem he solved by turning the cooling unit off. He also complained that while he'd been looking forward to experiencing the solitude of space, he had trouble doing so because of the three separate audio feeds to his earpiece. As he put it, "It is a true statement that sound does not travel through a vacuum. But radio waves do, and I had three different people talking to me. Mission Control wanted to know how much oxygen I had left, how the battery was doing, what was the temperature. Vance Brand, the commander, was wanting me to stay away from the engines and not go under the wing, stay where he could see me. And Bob Stewart wanted to know when it was his turn to fly!"

For McCandless and Stewart, the flight tests were prosaic—basic movement and maneuvering, things they'd done for hours at a time in simulation. For everyone else, though, the sight was extraordinary. Hoot Gibson grabbed the mission's camera to document the event. During the MMU's first flight on February 7, he looked through the viewfinder and was so amazed that he put the camera aside for a moment, unsure of what he was seeing. Then, fortunately, he focused again. As Chaikin has written, the photograph he took of Bruce McCandless operating the MMU "is one of the great mind-blowers of the 20th century." My father's old Naval Academy classmate, Senator John McCain, put it this way: "The iconic photo of Bruce soaring effortlessly in space has inspired generations of Americans to believe that there is no limit to human potential."

The Photograph is remarkable for a number of reasons, not least of which is its clarity. Taken with a NASA-modified Hasselblad camera outfitted with a 70-millimeter Zeiss lens and using ASA 64 Ektachrome film, the shot was composed manually—f-stop, focus, shutter speed—by Gibson. Technically, it's a flawless achievement, aided by the fact that the subject, reflecting the sunlight, appears to be glowing. The shot hints at the unfiltered vibrancy of colors as seen outside our atmosphere. It's also striking for the unexpectedness of the image—the astronaut is obviously unattached to any object and is apparently alone, high above Earth. As a consequence, the picture is hugely popular, consistently reported by NASA as one of its most-requested images. On social media the Photograph elicits visceral reactions. "My heart is pounding just looking at the pic," says one commenter on Facebook; "He is, hands down, the coolest dude that has ever lived," says another; and a third, more bluntly, notes, "Now THAT took some serious gonads!"

Aside from these surface matters, though, the Photograph stirs subliminal responses as well. The central figure is clearly human, even if helmeted and armored in Kevlar. And yet the figure floats—or, given its inclination, which implies movement, *flies*. This wasn't really a consequence of forward momentum, but rather of the astronaut orienting himself to the position of the orbiter, which was inclined some thirty degrees off horizontal (literally, the horizon of Earth, visible in the background). But the effect is the same. Bruce McCandless is, in a sense, italicized. We've always seen the ability to fly as wondrous. We've treated the act as a triumph over gravity, physical limitation, even mortality. Death means a return to dirt. Flight by contrast opens an avenue from earth to the eternal. In Christianity, Jesus rises from the dead and ascends into heaven to join the Father and sit in judgment of mankind. The Bible speaks of angels with wings, clad in dazzling white. The Greek god Apollo raced across the sky in a horse-drawn chariot with wheels of fire, and the Valkyries venture forth from Valhalla on winged steeds. The symbolic association of flight with the divine occurs in Eastern cultures too. When taikonaut Yang Liwei flew China's first manned spacecraft into orbit in 2003, he wore a shoulder patch depicting an *apsara*, a feminine spirit of the clouds, often shown airborne. Superman and Iron Man are pop culture cognates of the divine. Both accomplished fliers, they sacrifice themselves to save humanity from large, muscular, existential threats before the final credits roll.

But since real human beings are mortal, flight is inherently risky. We endanger ourselves when we imitate the gods. In Greek mythology, Icarus learns to fly but ignores his father's warnings and soars too high. The sun melts the wax that binds the feathers in his wings, and he falls to his death. Perhaps he should have bleached the wings. We've learned a few things about solar energy in the centuries since Icarus plunged into the Ionian Sea. The choice of white for the astronauts' pressure suits is practical, of course. White reflects radiation, including the heat of the sun. White is also highly visible against the backdrop of darkness that is outer space— important when trying to keep track of an astronaut. But the symbolic aspects of white in Western culture are also important to our understand-ing of the image. White is traditionally the color of purity and heroism, of the home team's jerseys, of the hat on the good cowboy's head. Thus, when we see the Photograph, we see ourselves, but better. Unfettered. Little pink and brown beings, creatures of mass and weight, regret and unease, now garbed in the robes of purity and free to leave earth. In effect, it's a tale of transcendence, told in high-tech terms. And since transcendence implies virtue, the Photograph is an emblem of spiritual victory as well, with the dreamy blues and whites and flitting angels of some medieval religious fresco reimagined for a skeptical century.

Then too there is a more conventional homily embedded in the shot. Many of the images we associate with the twentieth century are dispiriting, reflecting the era's slow dance with mechanized death. The fiery end of the *Hindenburg* in 1937. The ghostly dome of the Industrial Promotion Hall, still standing in August 1945 amid the radioactive rubble of Hiroshima. The chaotic, spidery swirl of smoke and flame radiating from the doomed orbiter *Challenger* in 1986 as its launch vehicle disintegrates around it. And then this: the oddly serene contrast of a solitary man emerging from the immensity of the universe, small but self-directed, neither obviously friendly nor immediately menacing, apparently in complete control. The image suggests order—a triumph, even if tenuous, against what is dark and immense and essentially incomprehensible. We *have* learned something, says the Photograph. We're moving. We're on our way.

WHILE HOOT GIBSON'S SHOT of Bruce McCandless against the backdrop of space has become a sort of visual shorthand for NASA and manned spaceflight, STS-41B was notable for accomplishments other than the successful test of the MMU. The crew performed experiments in the manufacture of latex beads, shot footage of the mission for Space 360 cameras, tested the effects of weightlessness on arthritic rats, and did experimental work designed by students at Utah State University and Scotland's University of Aberdeen. The mission made 127 orbits of Earth and traveled 2.8 million miles. On February 11 *Challenger* became the first orbiter to both take off from and land at the Kennedy Space Center when it glided in through a thin layer of ground fog for a picture-perfect touchdown. Previous shuttle missions had ended at Edwards Air Force Base in California. Landing in Florida instead—with its much shorter runway, surrounded by the so-called alligator moat—was an important goal for NASA, because it cut down significantly on the turnaround time required to prepare the orbiter for another flight. Still, it's difficult to ignore the legacy of the photographs—or, indeed, of the Photograph.

As was his custom, Bruce McCandless declined to indulge in hyperbole in describing his achievement. He was in all things, including spaceflight, relentlessly matter-of-fact. I'm sure my dad was elated to be gliding across the face of the planet. He just didn't think qualitatively. He was a quantity man. *How did it feel?* he was asked by a French interviewer. "My two 'spacewalks,'" he answered, "occurred on the fourth and sixth days of the mission. Consequently, I did have plenty of time to become accustomed to both the 'weightlessness' of space and to the beauty of the never-ending spectacle of the earth passing by below. We were at an altitude of about 330 kilometers; the 'scenery' appeared to go past at about five times the relative speed observed from a commercial airliner in cruising conditions. Having become accustomed to the scenery and to weightlessness, I was free to concentrate on the task of flying the MMU."

Resolutely levelheaded in public, Dad was more enthusiastic about the flight in private correspondence. One of his letters borders on giddiness: "In regard to David [Marrack]'s telegram, I cannot personally assert that the world is *spherical*, but it *is* round to the extent that we did not encounter any edges over which one could fall!" He called the views from the orbiter "wondrous" and "fantastic." He praised the hardware. "It's not often," he wrote, "that a complex piece of space hardware functions perfectly on its first trial, and [it's] practically unheard of for two units to behave identically." He expressed surprise that most of the public attention to his flight focused on its lack of tethers, and even engaged in a little technological preening. "We have regarded tethers as nuisances and potential hazards for quite some time. I should have been happier if people had been agog over the precise maneuvering capabilities demonstrated by the MMUs, or some similar feature." And he'd clearly been drinking the NASA Kool-Aid. "With the recent approval of a permanently manned space station," he said, "and the plans to start carrying 'citizen astronauts' in 1985 or '6, space is going to become a lot more accessible."

Singularly absent from Bruce McCandless's recounting of his exploits on STS-41B is a sense of fear. Did he feel it? He was an intelligent man. He knew the risks. Micrometeorites were a statistically slight but significant hazard; a random piece of debris could have sliced through his pressure suit or visor during an EVA and left him without oxygen and temperature control, killing him in moments. Gene Cernan almost passed out while trying to get to the AMU on Gemini 9. But staying inside the spaceship was no guarantee of safety either. Michael Collins and John Young found themselves flying *blind*—literally—for several minutes during Gemini 10 due to the effects of a small leak of lithium hydroxide in their oxygen supply. The American crew of the Apollo-Soyuz Test Project nearly died from inhaling toxic gases on reentering Earth's atmosphere. The orbiter *Columbia* actually caught fire shortly before landing just two months before STS-41B took off, though John Young was able to pilot the craft home safely. Four Soviet cosmonauts had been killed inside their capsules while returning from space missions, and three American astronauts were killed while preparing for one.

So the risks, known and unknown, were there, but my dad had had some experience with fear. He'd landed Phantoms on the deck of an aircraft carrier at night, after all, with the Redheaded Gunman watching from the end of the runway. In his view, there was nothing more dangerous. Then, too, he wasn't just *flying* the MMU. He'd spent eighteen years with some of the best technical minds of his generation, testing and developing the device. To have admitted fear about its operation would have called into question his engineering skills, which were formidable. He was ready and more than willing to fly. In fact, he and fellow astronaut Bob Stewart had to argue with NASA management to avoid the ignominy of having to perform their tests of the MMU while *tethered*. Chaikin tells the story:

> Back in 1966, worried that Gene Cernan could end up a permanent Earth satellite, managers had insisted the AMU be operated with the astronaut attached to a 75-foot lifeline. Now, for the MMU, they were talking about a tether of 300 or more feet. According to McCandless, NASA spent $3 million to develop what looked like an oversize fishing reel mounted on the front of a space suit. If the jetpack conked out, its pilot could simply crank a handle and pull himself back to the payload bay.

My dad and Stewart simply refused to test the MMU while attached to the shuttle. In a rare win for astronauts over the bureaucracy, they finally got their way. So were they scared when the time came to leave the orbiter and head out into space, with only a limited supply of nitrogen gas between them and eternity? Did my father think back to those moments when he was tumbling violently through the sky above the Mediterranean in

1961 as his plane plunged into the wine-dark sea? I doubt it. I never saw him scared. And it didn't matter. Even if he'd been scared, it wouldn't have stopped him. And yet even heroes are human. My dad has noted that he was supposed to head out 300 feet from the orbiter and *turn around*, facing away from the spacecraft, the only way he had to get back to Earth. He wasn't afraid that the MMU wouldn't work, or that he would have some mental lapse, or that the shuttle itself might start moving away. None of that was in the script. But even he later admitted that despite his intentions, he never turned his back to *Challenger*.

WHEN HE MANEUVERED AWAY from his crewmates on the shuttle, which was in orbit 178 miles above Earth, Bruce McCandless was—like the orbiter—traveling at a speed of 17,500 miles per hour relative to the planet. This is an approximation often used as shorthand for the speed a spacecraft in Earth orbit has to maintain in order to avoid either soaring off into outer space (if its orbital velocity is too fast) or eventually sinking back into the atmosphere (if its orbital speed is too slow). It's worth noting that this velocity is twenty-three times the speed of sound. After a brief test, my father ventured out to 320 feet away from *Challenger*—still an EVA record—on the power of the MMU's modest, 1.4-pound nitrogen thrusters. He made a total of five sorties over the course of two days, logging between three and a half to four hours of flight time. But it was his initial flight on February 7, 1984, that guaranteed him a spot in the pantheon of space explorers as the world's first human satellite.

Plaudits poured in. In the months following the return to Earth of STS-41B, he did interviews with domestic and international journalists. Washington sent him on a goodwill tour to Africa, including a stop in Zambia to meet with President Kenneth Kaunda. He was a celebrity contestant on the game show *What's My Line?*, on which he won $333.34. Maybe more importantly, his accomplishment was lauded by his fellow astronauts. Bill Anders, for example, who'd journeyed to the moon and back with Frank Borman and Jim Lovell in 1968, wrote with words of praise. The first flight of the MMU, he opined, ranked in importance with the Mercury exploits of Shepard and Glenn, Apollo missions 8 (Anders's flight) and 11, and the first flight of the space shuttle. Anders also acknowledged the "long road" my dad had traveled before his mission. After eighteen years, the Boy Wonder who'd fallen from grace had finally pulled himself back up again. He'd made it into space at last. He could exhale.

THE MMU CAPTURED THE public imagination. It's a machine people like to believe in. In the 1969 movie *Marooned*, for example, Richard Crenna, Gene Hackman, and James Franciscus star as the crew of an Apollo command module that docks with a Skylab-like space station to commence an extended stay. Their first task is testing the "S-509," a jetpack maneuvering unit. After leaving the station several weeks later, the crew is stranded in space when their command module's engines fail, leaving them unable to initiate reentry. The astronauts' oxygen supply runs low, and one of the crew dies in an effort to keep his comrades alive. As time runs out for the remaining two crewmembers, a Soviet *Soyuz* capsule arrives on the scene to attempt a rescue. A plucky Slavic cosmonaut emerges from the capsule and flashes a series of hand signals, roughly translatable as, "Why does everyone move so slowly in this movie?" The communist spacefarer is sadly hampered in his rescue efforts by the fact that he only has what appears to be a Gemini-surplus handheld maneuvering unit with which to translate over to the stricken command module. James Franciscus tries to help by propelling the unconscious Gene Hackman toward the *Sputnik*. He miscalculates, however, and instead throws Hackman in the general direction of Micronesia, like a giant dart with feet. Fortunately, David Janssen, playing a thinly disguised Deke Slayton, shows up just in time in what looks like Hollywood's version of the X-24A space plane. Unlike the technologically challenged cosmonaut, Janssen has a manned maneuvering unit, which he uses to rescue the two apoxic astronauts before returning, triumphantly, to Earth.

Other cameos: American space marines use manned maneuvering units and laser pistols to capture supervillain Hugo Drax's orbital lair in the 1979 James Bond flick *Moonraker*. Jessica Chastain dons an MMU-like jet pack to rescue Matt Damon in the 2015 film *The Martian*, and George Clooney flies a souped-up version of the device to save Sandra Bullock in 2017's *Gravity*. Sadly, however, the real MMU's functional life was short. Despite its successful—and spectacular—test flights, the machine saw only limited use after STS-41B. It was used unsuccessfully on STS-41C in April 1984, when George "Pinky" Nelson flew out to the malfunctioning Solar Max satellite and attempted to dock with it in order to bring the satellite back to the shuttle. The jetpack worked fine. Unfortunately, the docking mechanism Nelson used was unable to properly connect with the satellite due to the presence on the satellite of a tiny metal grommet the manufacturer had neglected to note in its blueprints. Frustrated, Nelson tried grappling the satellite by hand but only made the situation worse. In the end, he returned to the shuttle with nothing to show for his efforts. Technicians on the ground were later able to work with astronauts on the shuttle to slow the satellite's rotation and use the robotic arm to bring it in to the orbiter's payload bay, where it was successfully repaired and redeployed. The MMU was more helpful on STS-51A later that year, when Dale Gardner and Joe Allen were able to use the machine to retrieve the two malfunctioning communications satellites deployed by STS-41B and wrestle them into the shuttle's payload bay for return to Earth. Photographs of Gardner pushing the much larger Westar VI through space, like an ant with a giant crumb, are some of the most bizarre and beautiful of the shuttle era. In each of these cases, the MMU performed as planned.

Ultimately the MMU's demise was occasioned not by any design or performance problem, but because NASA figured out that much of what the astronauts needed to do during the shuttle era was doable by more conventional means, through use of the robotic arm and tethered spacewalks and with the aid of the very maneuverable orbiter. After the *Challenger* disaster in 1986, the MMU simply seemed too risky compared to these alternatives. There is a certain amount of irony in this position. It's unclear why NASA would have discontinued a demonstrably viable machine like the MMU due to some hypothetical danger at the same time the agency was hell-bent on flying shuttles in the face of well-documented risks like O-ring shortcomings and foam-insulation strikes.

Bruce McCandless, Ed Whitsett, and Martin Marietta lobbied hard for additional work for the MMU, but to no avail. There was intermittent interest from the agency and from other government entities. In March 1988, for example, my father wrote in his journal: "[Whitsett] sez there is DOD/USAF interest in a MMU mission in Aug '88! [Name withheld] prohibited him from discussing it with me . . . It's something he never would have imagined." The timing almost matches up with an enduring shuttle-era mystery: namely, how was the crew of STS-27, which flew in December of 1988, able to affect repairs to a faulty top-secret Defense Department satellite the mission deployed? The satellite's antenna dish failed to open, which threatened to turn it into a floating piece of space junk. STS-27 crewmembers have mentioned having to make repairs to the satellite, but have never said how this was accomplished. Many observers think an EVA was performed. Some suspect an MMU may have been used. For now, though, the mystery continues, as the mission was classified, and neither NASA nor the Pentagon has ever been willing to discuss it. At any rate, according to Bruce McCandless's journal, DOD's interest in the MMU—tantalizing and hyperbolic—dissipated shortly after it was expressed. The three MMUs Martin Marietta produced are now display pieces. Unit 3, the first to fly, and the machine shown in Hoot Gibson's famous photograph, hangs above the orbiter *Discovery* at the National Air and Space Museum in Washington, D.C. Unit 2 resides in Huntsville, Alabama's Space & Rocket Museum. And Unit 1, which never flew, remains at the Johnson Space Center.

While the MMU is in mothballs, a smaller, lighter version of the jetpack was eventually produced as SAFER—the Simplified Aid for EVA Rescue device—and is still used by American astronauts on all EVAs to this day. Weighing in at 85 pounds, SAFER fits around the Extravehicular Mobility Unit backpack and, like the MMU, utilizes twenty-four nitrogen-gas thrusters to provide six-degrees-of-freedom maneuvering capability. The propulsion system generates a small but appreciable total delta velocity of at least ten feet per second with an initial charge. It is actually more of a life jacket than a transportation device. NASA's 1993 SAFER training manual describes the device as a "small, self-contained, propulsive backpack system to provide free-flying mobility for an EVA crewmember. SAFER is a 'simplified' single-string system for contingency use only, SAFER does not have the propellant capacity and systems redundancy that was provided with the MMU." In other words, SAFER is to the MMU what the Smart Car is to a Corvette. Though the SAFER was successfully tested on shuttle missions STS-64 (in 1994, with astronauts Mark Lee and Carl Meade untethered) and STS-92 (in 2000, with astronauts Peter Wisoff and Michael Lopez-Alegria tethered to the orbiter), to date it has not—thankfully—been used in an emergency.

It's certainly possible to imagine a future for humanity in space without jetpacks. But why would anyone want to? Robots and uncrewed probes are increasingly sophisticated and versatile tools for sampling the solar system. Still, it's too early to surrender all the joys of movement and curiosity to automation. Bruce McCandless felt that the MMU would be the perfect way to maneuver around the irregular surfaces of asteroids. The jetpack could be equipped with hardware for attachment to exploratory spacecraft and hauled along for reconnaissance use. I think my dad's suggestion in this regard was intended as a hint to NASA brass. If so, let's hope they heard it. If not, someone else will. Many of us are interested in space for its scientific allure. But we may be in the minority. If the past is any guide, the cosmos will eventually be exploited for commercial gain or occupied for military advantage. The two go hand in hand. Scientific curiosity is a persistent spur to exploration, but commercial and national security interests often provide a more powerful push. In these circumstances, the utility of the jetpack will outweigh its supposed risks, and some version of the machine doubtless will fly again, for rescue, repair, or possibly even recreational purposes. The MMU was, as Pinky Nelson put it, a beautiful example of aerospace engineering. Bob Stewart said it was the easiest thing he'd ever flown. "The only way you could make it easier," he added, "would be to wire it directly to your brain."

15

INTERMISSION

M y college experience was a little different than my father's. There was the arrest, for example. It was just after noon on March 2, 1982, when the cops put me in the squad car. Did I understand that I was under arrest? one asked. Sure, I said. I had a pretty good idea. The 200 UT students around me certainly understood, and they weren't happy about it. "The PEOPLE!" they chanted. "UNITED! Will NEVER be DEFEATED!"

It was student politics in the age of Reagan. I won't bore you with the details of this particular fight for truth, justice, and the American way. We lost. I only mention the incident by way of saying I couldn't always relate to the parade ground spit and polish of the Naval Academy. After high school, I loaded all my belongings in my VW Super Beetle and headed off to the University of Texas. I wasn't notably self-reliant growing up, preferring to have my mother perform all but the most vital life functions for me, but I'd gotten it in my head that I was going to pay my own way through college. While this meant that I never had any money, it didn't much matter. Austin was an oasis for the impoverished. Mrs. Nguyen and her extended family sold eggrolls for fifty cents apiece on the Drag, and a local enclave of saffron-robed Hare Krishnas fed all comers a meal of rice and steamed vegetables for the price of only a little interest, simulated or otherwise, in the avatars of Vishnu. Sister Cindy and the Reverend Jed preached on a more or less continual basis at the foot of the West Mall, and you could watch them haranguing the girls from Tarrytown Methodist for hours at a time. You could wander up to the fourth floor of the UGLy, the under-graduate library, and listen to recordings of Django Reinhardt and Blind Willie McTell all day if you wanted. Weekends were even better. On any given Friday there were a dozen movies playing around campus for a dollar a show. Howard Hawks and Ingmar Bergman, Harold Lloyd and Jean-Paul Belmondo: only a dollar. On Saturdays a group of us would head east of I-35, the de facto ethnic and racial dividing line of the city back then, and eat *migas* with tortillas and refried beans at El Azteca for a couple of dollars a plate.

In April, Barton Creek came tumbling down from the hills west of town toward the crystalline pool at Barton Springs, and it didn't cost a dime to sun yourself on one of those slabs of limestone beside Twin Falls and watch idiots go shrieking over the drop. As the weather warmed, there were free festivals and concerts, a Spam cook-off, a citywide tug o' war, footraces, pun-offs, and a massively stoned celebration of Eeyore's birthday. The next Stevie Ray Vaughan lived in every van. Frisbees freckled the sky, and sometimes the city seemed like a massive conspiracy mounted by those who didn't have any cash to annoy the people who did.

IT WAS AN ODD time to be in Austin. My friends and I missed the city's glory days, a semi-mythical decade or so when landlords declined to collect the rent, Janis Joplin lived next door, and Willie Nelson would show up in your living room to play "Whiskey River" every Fourth of July. Still, some of the magic endured. Sometimes it felt like we were swimming in a temporal eddy of the national tide. The country's political mood was a sort of rejuvenated jingoism, a sense that Johnson and Nixon and Carter had allowed a decline in the nation's global status that Ronald Reagan was going to put right. Reagan had followers at UT, but the dominant senti-ment on the West Mall was cynicism. Unlike our parents, the children of the fifties, some of us suspected America was ugly inside. We'd seen the footage from the fall of Saigon. We'd read about the death of Joe Campos

Torres, a Vietnam vet who'd been handcuffed, beaten by Houston police, and thrown from a bridge into Buffalo Bayou. We wondered who really shot JFK. Unlike Reagan's acolytes, we'd seen the enemy, and it was us.

The Clash sang "I'm So Bored with the U.S.A.," and Bruce Springsteen released *Nebraska*, an extended paean to lost lives, drifters, and American failures. On the Drag, just west of campus, a little club called Raoul's hosted a rotating assembly of punks with knock-off Telecasters—the Skunks, the Huns, the Big Boys—who rehearsed the end of everything on a nightly basis. When student government returned to campus in 1982 after a hiatus of several years, UT students elected a cartoon character named Hank the Hallucination to the newly resurrected position of student body president, before a young liberal named Paul Begala claimed the win on a technicality. (Being a cartoon character, Hank couldn't actually raise his right hand to take the oath of office.)

Despite the political ennui on campus, Austin was busily reinventing itself. John Mackey opened up his first Whole Foods Market in a one-story building at 10th and Lamar, and Michael Dell started cobbling computers together in his dorm room in Dobie Center. Louis Black and Nick Barbaro, publishers of a free alternative newspaper called the *Austin Chronicle*, dreamed about starting the musical festival that would become SXSW. Future faces in the arts, science, and politics like Marcia Gay Harden, Neil deGrasse Tyson, Richard Garriott, Berke Breathed, and Sam Hurt all walked the Forty Acres. They were soon followed by Wes Anderson, Richard Linklater, Robert Rodriguez, Renée Zellweger, and Matthew McConaughey. Tuition was cheap. Eccentricity was admired. The city and the college drank and danced with each other, and a new Austin has been born every year or so since. It keeps getting bigger and faster and brighter. Its resemblance to its parents fades with every generation.

My dad got it. He may have been disappointed I didn't follow his path to the Naval Academy, but one look at my car, with its peace-sign bumper sticker and its foot-deep detritus of movie posters and fast-food bags, suggested maybe military life wouldn't be a great fit. I agreed. If studying literature at UT wasn't exactly the path to enlightenment, enrichment, and maturity, it nevertheless seemed infinitely preferable to doing push-ups on the linoleum for some crew-cut midshipman with a Nimitz complex and a drawer full of Boy Scout merit badges.

I lived in sweaty poverty those days with a motley band of petty thieves and library clerks, strangers to any sense of decorum or decency. We occupied a cement house at the far end of Austin's West 22nd Street that made Bancroft Hall, the Naval Academy's massive dormitory, seem posh by comparison. We had one bathroom for five dudes and no air-conditioning, which in central Texas ranks somewhere between stupid and suicidal. My mom and dad visited to inspect my living accommodations and reported back to friends and family with evident glee. For the rest of his life, my father enjoyed telling people about how I suspended a bag of potatoes from my ceiling so the rats wouldn't eat them before I could. He seemed bemused by my love of fiction and film, at least to a point. The one time I saw him flinch at a mention of my undergraduate pursuits was when I informed him, in the fall of my junior year, that I'd seen twenty-six movies in the first seven weeks of the semester. Twenty-six and a *half*, I corrected myself, as I hadn't been able to sit through Werner Herzog's *Angst Essen Seele Auf.* Dad was unimpressed by the precision of my accounting.

I graduated from UT in May 1983 and went to England on a Rotary Foundation fellowship, a generous award that required only that I occasionally give talks to local Rotary clubs about life in the United States and my impressions of the United Kingdom. Dad drove me to Houston Intercontinental Airport for the first leg of my trip on a scorching hot afternoon in September 1983. Though he was deep in training for his first shuttle flight, he insisted on delivering me to the terminal himself. He never said as much, but I could tell he was excited. For one thing, he talked as he drove. He acknowledged that I was an adult now, more or less, and advised me to stay away from the women I might meet in places like Piccadilly Circus, which was, or had been at one time, a known haunt of prostitutes. "Prostitutes," I murmured. "You mean, like . . . ?" He nodded grimly. I assured him I would be cautious with the women I met in Piccadilly Circus. I flinched to think what might be coming next. *You know, son, when two spacecraft find*

themselves in the same or similar orbital trajectories, and the decision is made to
commence docking, there are certain precautions that need to be . . . Fortunately,
we were out of time. We unloaded the trunk from the Volvo, the same trunk
I'd taken to camp all those years ago and exchanged the customary blood-
less McCandless handshake.

"Good luck out there," I told him.

I was a chip off the old block.

WHILE MY DAD'S FREE flight in space had delivered a supernova-size
jolt of euphoria, the period after his return to Earth resulted—as it had and
still does with other astronauts—in a long, slow period of disaffection. In
September 1984 he was still riding high. Hoot Gibson's photographs of my
dad flying the MMU were a sensation, so good that they seemed to reinvent
the occasion as something less like science and more like art. NASA was
working hard to keep its promise of regular launches, with lots of satellite
launches and servicing prospects, and the MMU was performing flawlessly.

I had returned from overseas at this point, and my father and I traveled to Nebraska that month for the dedication of the Rock Creek Station
State Historical Park. Located just south of Fairbury in the southeast corner of the state, the park was created on the site of the toll bridge and
Pony Express station where Dave McCanles was gunned down by Wild
Bill Hickok. The old station stood beside the Oregon Trail, and as my
dad spoke to the crowd I could see wagon wheel ruts in the earth, leading
west. He compared the space shuttle to those old Conestoga wagons, both
instruments of exploration and commercial enterprise, and remembered
the pioneers as exemplars of the American spirit. It wasn't one of his better
speeches. The whole shuttle/covered wagon thing was a stretch. And yet he
was excited to be there, and I knew the reason why. It was his appreciation
for the historians and park personnel who had worked to uncover the truth
about the Rock Creek shooting 120 years earlier.

Hickok hadn't, as some maintained, killed ten bloodthirsty members of the "McCanles Gang" on that long ago afternoon. Rather, he'd shot and killed Dave McCanles in cold blood, because McCanles was trying to collect an overdue debt from Hickok's boss, who managed the Pony Express station. The murder left McCanles's wife a widow and robbed his children, including a boy named Julius (my dad's great-grandfather) of a father. The Rock Creek Station ceremony, attended by Governor Bob Kerry and a gaggle of McCandless kinfolk, helped to mend some old wounds. It was simultaneously a repudiation of Hickok and his lawless ways and a vindication of the McCandless name. For a man as concerned to honor his forebears as my father was, it was a tangible triumph. For one summer afternoon in Nebraska, the Redheaded Gunman and all the ill omens he carried with him were nowhere to be seen.

By 1985, though, Bruce McCandless found himself wondering whether to leave NASA. Robert Reinhold of *The New York Times* suggested that the graying Buck Rogers was on his way out, and Dad does seem to have been assessing his options. The work was still intriguing. With Head of System Engineering Jack Keeney and others at the Boeing Company Space Division, he began training to use the Inertial Upper Stage (IUS)–installed hardware in the shuttle. The IUS was a two-stage solid rocket booster for satellites intended for much higher orbits than the shuttle orbiter could achieve. Its first use on a shuttle mission was on STS-6 in 1983, when it placed a tracking data relay satellite in geostationary orbit. Later, a third stage was added to the IUS to boost interplanetary probes—the Magellan probe to Venus in 1988, the Ulysses solar pole mission in 1990—and, in 1999, to launch the Chandra X-ray observatory.

But like any astronaut, my father fretted about his prospects of getting back into space. In one journal he kept clippings related to the retirement of a fellow astronaut. Notes memorialized a phone call he had with the colleague: "No future in being an MS [Mission Specialist]—if you're good, you get to be an MS again—limited horizon." Plus: "with flights getting closer together, the CDRs [commanders] have more time to meddle in the MS's responsibility area." Another retiree he spoke with said that "the

day on which he walked his retirement request through the chain of command—Young—Abbey—Charleston—Moore—he felt like he was finally
his own master. Expressed resentment at the power exercised by GWSA
[George Abbey] over such a broad area."

In April 1985 Senator Jake Garn of Utah flew on *Discovery* as a payload
specialist for STS-51D. While Bruce McCandless had nothing against the
senator personally, he found it galling that a politician was taking flight
time away from qualified astronauts. He kept a mock NASA memo from
that year in which the top brass announced that "they will conduct a selection from among the 92 active astronauts to find a replacement to fill Sen.
Garn's vacancy while he undergoes NASA training." At least Garn was a
former Navy pilot. NASA next signaled that it was interested in putting
civilians of all stripes in the shuttle. Walter Cronkite applied for a spot.
John Denver was interested, too, and willing to pay, but ultimately President
Reagan determined that a teacher should be the first nonpolitical civilian to
get a place on the space plane. It was a smart decision. We all loved Christa
McAuliffe. The agency's well-publicized search to put an educator in the
cosmos seemed to have found exactly the right person. A spunky, unpreten-
tious thirty-seven-year-old from New Hampshire, McAuliffe was lauded
by neighbors and students as relentlessly positive, curious, and energetic.
Like Ripley in the movie *Alien*, even down to her shoulder-length perm,
she was feminine but sturdy, with a determined set to her jaw when she
wasn't smiling, as if she could handle a pulse rifle as well as a paring knife.
We knew she was nervous. Anyone would have been. But she made it clear
she'd be fine. She was confident in the skills of her commander, Dick
Scobee, and her crewmates, pilot Mike Smith, mission specialists Ron
McNair, Judy Resnik, and Ellison Onizuka, and payload specialist Gregory
Jarvis. She knew she could rely on the expertise of NASA management.

And so on the day President Reagan was scheduled to give his fifth State
of the Union address, *Challenger* surged up off Launchpad 39-B on a cold
January morning and flew for a little over a minute. That's how long it took
for malfunctioning O-rings to let superhot gas from one of the solid rocket
boosters cut a hole in the orbiter's main fuel tank. The shuttle broke apart

at an elevation of 48,000 feet, almost nine miles up, in a chaotic tangle of divergent vapor trails. It was an image unlike any we'd ever seen before, an irrational snarl that is nevertheless instantly recognizable today to anyone who saw it happen. Parts of the orbiter, including the crew cabin, continued to ascend for perhaps as many as twenty-five seconds afterward, even as the shuttle's rocket boosters snaked away, bereft of guidance from the shuttle's computers. Inexorably, the cabin's ascent slowed. And then it fell.

Parts of *Challenger* were never found. A lot of what was located came to rest on the floor of the Atlantic Ocean, a hundred feet below the surface. Suddenly the question of who was going to fly on which mission became trivial. The president's Rogers Commission, a panel that included such talented but diverse individuals as Neil Armstrong, Sally Ride, and Richard Feynman, concluded its investigation of the disaster with a searing indictment of institutional arrogance at the agency. NASA suspended all shuttle missions for a period that ended up being more than two and a half years. It was a stunning setback for the agency. There had been space-related fatalities before, of course: the astronauts of Apollo 1; Vladimir Komarov on Soyuz 1; the crew of Soyuz 11. But the most recent of these accidents had occurred fifteen years earlier, and none had been broadcast live on network television, either here or abroad, on a day when NASA felt so convinced of its own excellence that it ignored clear warnings about the dangers of a cold-weather launch.

"Early 1986, boom," said one astronaut, looking back. "Here comes the *Challenger* accident, and it just turns NASA upside down. Never before that had anybody thought that NASA was less than the best-managed federal agency ever. Immediately after, we found that we were the worst federal agency ever. You know, the tables just turned around completely."

Bruce McCandless was already restless. Now his dissatisfaction blended with disappointment and sorrow for the loss of his colleagues, particularly his friend Ron McNair. He even penned what sounds like the draft of a defiant resignation letter: "If the orbiter is destined to turn itself into a smoldering pile of scrap with no opportunity for my intervention, I see no particular reason to add my body to the debris." It's hard to know whether he was referring here to the safety problems of the shuttle generally or in particular to the lack of any means of escape or ejection from a troubled orbiter. He may also have been reflecting on the fact that in his role as mission specialist, rather than a commander or pilot, he wouldn't have been able to correct any in-flight problems that developed on a future mission. Though he'd been willing to fly on STS-41B as a mission specialist rather than as a pilot, and in fact would have been willing to fly if wrapped in Kevlar and strapped into the payload bay, he always hoped for a front-seat flight role as a commander or pilot. This wasn't going to happen with his recent assignment to an intriguing but long-delayed project, the Large Space Telescope. The project, which would eventually be known as the Hubble Space Telescope, was already way over budget and several years late in development. An impatient Congress had threatened to zero out its funding at one point. The Challenger disaster set the time line for launch back even further.

So Bruce McCandless had an important decision to make. Leave the manned space program and cash in on his new image as the world's first human satellite for a cushy and risk-free job with one of the aerospace contractors? Or stay where he was and shepherd the space telescope—an expensive and iffy piece of scientific machinery—into Earth orbit aboard a shuttle that he now knew to be much more dangerous than originally billed?

REFLECTING HIS INNER DELIBERATIONS, Bruce McCandless continued to explore his options to what life as an astronaut had become in the eighties: less glamorous but just as demanding, more bureaucratic but equally, if not more, dangerous. He enrolled in the MBA program at nearby University of Houston–Clear Lake. He read and dutifully answered questions in a workbook called *The Inventurers*, a sort of *What Color Is My Parachute?* for late-in-life job hoppers. He attended a career-planning seminar and, apparently as an exercise related to it, wrote and mailed himself a letter setting out goals for personal development: start a business; socialize more; spend time traveling with family. But during this post-*Challenger* period, when the temptations to leave NASA seemed most compelling, he found himself fascinated by a side project that was equal parts entertainment and engineering.

Before *Challenger* went down, NASA had assigned Bruce McCandless and his STS-41B crewmate Ron McNair to work with French composer Jean-Michel Jarre on the musician's plan to create an outdoor multimedia concert celebrating the 150th anniversary of Houston's founding. Part of the presentation would depict the evolution of NASA and the Johnson Space Center—hence the agency's interest.

Long-haired and unshaven, with a glass of wine in one hand and the actress Charlotte Rampling holding the other, Jarre was a phenomenally popular composer of electronic music at the height of his powers. His charisma cast a spell on the normally staid astronauts, and they became enthusiastic co-conspirators. Indeed, on STS-41B, McNair recorded a saxophone solo for use in the concert, to mixed reviews. For one thing, it was hard to play in the reduced-oxygen environment of the orbiter. For another, there was leakage. "Looks like spit trap on this thing doesn't work in microgravity," McNair commented afterward. "No," his crewmates affirmed. "It doesn't."

After the *Challenger* disaster, Jarre was tempted to cancel the production out of respect for the deceased. My father urged him not to. Continuing with the concert, he said, would be the best way to honor the crew. It's unclear how much influence Bruce McCandless's words had, but the concert, called Rendez-Vous Houston, did in fact take place on April 5, 1986. It was big, bizarre, and beautiful. Traffic on downtown streets and freeways came to a standstill. Drivers left their cars and sat on the pavement to watch. As Jarre played his New Age music on what looked like a large flying saucer, projectors flashed video clips and static images—a dove, the state of Texas, the Manned Maneuvering Unit—on the steel-and-glass faces of giant skyscrapers. Lasers glowed. Fireworks flowered in the sky and fountained off bank building roofs. At one point the heavens were so filled with fire it looked as if a race of synthesizer-loving extraterrestrials had journeyed to Earth to blow themselves up over Buffalo Bayou in a fit of sensory ecstasy. Jarre abandoned the flying saucer to play notes on a lawn sprinkler that sprayed beams of light. A boys' choir chimed in to lend an air of celestial sanction. Jarre next stood in a tube of blue phosphorescence, as if encased in a neon cigarette. And now Ron McNair appeared in a video on the central screen. He somersaulted in the belly of NASA's zero-g training plane and jogged along a beach at sunset as Kirk Whalum, standing alone on a raised platform, eased into a moody extended saxophone solo called "Ron's Piece."

Something like a million and a half Houstonians saw the performance that evening, a new record for an outdoor concert. And then all went silent, and downtown was dark until a black-and-white video of John F. Kennedy flickered to life on the biggest building. It was the dashing, doomed president speaking on a long-ago afternoon at Rice University, challenging the nation to put a man on the moon and bring him safely home by the end of the decade. "We choose to go to the moon," he said, "not because it is easy, but because it is *hard*." The cheer from those gathered for the concert— old and young, black and white—was deafening. Defiant. Undaunted. Rendez-Vous was a summons to the stars: a farewell, a tribute, and a rededication all at once.

In my view, Bruce McCandless's involvement in Rendez-Vous Houston reminded him that even in the aftermath of failure, his passion for space exploration was shared by French composers, South Carolinian polymaths, and the hundreds of thousands of ordinary people who stopped their cars on the freeway to watch images of the future flash and fade on the reflective screens of Houston's skyline. The adventure had paused, he realized, but it was far from over. He made up his mind.

He was staying.

16

LAUNCHING
THE TIME MACHINE

It had to be *Discovery*. Think about it. There were three active orbiters in 1990: *Atlantis*, *Columbia*, and *Discovery*. Only one of those names really fit for deployment of the most important piece of scientific hardware ever produced, the Hubble Space Telescope. Now, after all these years, it seems hard to recall a period when the satellite wasn't up there, circling Earth, beaming down images and information that regularly reorder our understanding of not only our solar system but also our galaxy and, let's face it, everything else. And yet its launch and long life were not at all predictable at the time of its development.

One of the great early rocket scientists, Hermann Oberth, dreamed up the idea of sending an observatory into space. The American astronomer Lyman Spitzer apparently initiated serious discussion of the benefits to be derived from such an instrument with a paper he published in 1946. First and foremost among such benefits, he wrote, was the ability of an orbital telescope to make observations unimpeded by the various distortions caused by our atmosphere, which is a chaotic soup of gas and dust. This soup blurs visible light, causing stars to twinkle and making it difficult to see our neighboring planets and more distant objects. It also hinders or even totally absorbs other electromagnetic wavelengths, making observations in such wavelength ranges as infrared, ultraviolet, gamma rays, and X-rays difficult or virtually impossible. Spitzer became a tireless advocate for the idea of a

space observatory and over the years managed to convince many of his colleagues of the value of such a tool. They in turn lobbied their lawmakers to
support the vision. Congress was not so easily swayed. But after several fits
and starts, construction of the space telescope commenced in 1979 at Lockheed's facility in Sunnyvale, California. Originally scheduled for launch in
1983, the mission was repeatedly delayed due to development problems. A
subsequent 1986 launch date for a shuttle deployment mission commanded
by John Young was postponed when *Challenger* was destroyed.

An athlete as well as a thinker, Edwin Hubble was a Rhodes Scholar,
army officer, boxer, and lawyer whose interest in the scientific aspects of
astronomy eventually lured him into a career in astrophysics. It was Hubble,
working from the Hooker Telescope at Pasadena's Mount Wilson Observatory in the 1920s, who demonstrated that nebulae, the cloudlike images
visible in the night sky, are in fact galaxies beyond our own Milky Way, each
consisting of millions of stars. Later, using shifts in the light spectra emitted by stars in these nebulae, he also laid the groundwork for the discovery
that other galaxies are moving away from ours and that the galaxies farthest
away are moving at the greatest relative velocities. The significance of his
work made the pipe-smoking, collegial Hubble, a son of the Midwest, a
logical choice for commemoration. In 1983 the Large Space Telescope was
officially renamed for Edwin Hubble.

Like any number of space projects, Hubble, a joint undertaking of
NASA and the European Space Agency, was not quite as grand a machine
as its designers originally envisioned. Even so, the final product was a magnificent achievement—what some stargazers have described as the biggest
technical improvement in astronomy since Galileo fashioned his spyglass
in 1609. Fifty feet long and thirteen wide, weighing in at a little over twelve
tons, the satellite is often likened in size to a school bus. In space it looks
like a giant telephoto lens wrapped in aluminum foil, with the business end
of a rubber spatula attached as a lens cover. Its main mirror has a diameter
of 2.4 meters. Dr. Holland Ford, an astronomer affiliated with the University of Wisconsin and the Space Telescope Science Institute, calculated
that Hubble could detect a firefly as far away as the moon. Further, if there

were *two* bugs, and they were more than nine feet apart, Hubble could tell if there were *two* fireflies or just one really bright one. One report said the satellite, which ultimately cost something like $1.5 billion to develop, would be capable of distinguishing the period at the end of a typewritten sentence from a mile away.

The point wasn't periods, though. The Pentagon asked, confidentially, if the new space observatory could be used to study objects in geosynchronous orbit—like, say, other nations' satellites. Bruce McCandless responded politely that, considering Hubble's tracking rate, focus, and resolving capabilities, the observatory probably wasn't a good candidate for such relatively close-up viewing. Scientists, meanwhile, were looking for clearer pictures of any number of objects and occurrences beyond Earth orbit. Light travels fast, more than 186,000 miles per second, but it still has to travel. The sunlight we enjoy on an April afternoon left our local star eight minutes and twenty seconds before it reached our faces. By gazing deep into space, Hubble could in effect look back into time itself, and thus detect what the universe looked like billions of years ago.

BRUCE MCCANDLESS STARTED WORKING on Hubble in 1978. He and Kathy Sullivan were assigned to the Hubble deployment mission as mission specialists late in 1984. An oceanographer by training, Sullivan was only thirty-three, but she had already claimed the distinction of being the first American woman to walk in space. Tall and physically fit, Sullivan was always a tomboy. She combined an impressive intellect with a direct manner and was perhaps the most articulate member of the large astronaut class of 1978. Years later, in 2020, she would also become the first woman to visit Challenger Deep in the Western Pacific, the deepest known point in the world's oceans.

Originally the plan for prolonging the Hubble Space Telescope's functional life was to retrieve and return it to Earth via the shuttle periodically for maintenance and repairs. Somewhere along the line, as it became clear that the shuttle was not going to be quite as frequent or as cheap a flier as envisioned, a decision was made that Hubble would need to be fully maintainable in space, that is, *while it was in orbit*. This was a revolutionary concept, and the practical problems were formidable. Think of it this way. You and your fellow taxpayers have just paid more than a billion dollars—more than has ever been paid to create any scientific apparatus in the history of the world—to manufacture an innovative, highly sensitive piece of machinery that scientists hope will give humankind an unprecedented understanding of its place in the cosmos. If anything goes wrong, the machine will have to be fixed where it's found, and you're part of a team trying to figure out how this will happen, given that Hubble will be 300-plus miles above the Earth, in an environment that will kill any living thing that isn't wearing a pressure suit, and any tool you use will go floating away into the cosmos the moment you let go of it. Add to these challenges the problems inherent in tinkering with anything that's been designed so that lots of important parts fit in a very small space. As Sullivan has said, working on Hubble's main power unit is a little like trying to change the spark plugs in your car while wearing two snowmobile suits, a pair of mittens, and a bucket over your head.

For the next several years, including those bad, blue days after the *Challenger* accident, McCandless and Sullivan spent countless hours with engineers and technicians from NASA and the Lockheed Corporation, planning and practicing how to make maintainability happen. A NASA engineer named Jean Olivier dreamed up a toroidal architecture for the telescope in which systems were housed in modular units around the body of the satellite. A great start, but how, exactly, to get at and into these units? The Hubble Space Telescope isn't actually a single device but rather a collection of instruments contained in that giant, lens-shaped can. At its heart the cylindrical optical telescope assembly holds a Cassegrain reflecting telescope, in which a 2.4-meter-diameter primary mirror first receives the light from astronomical objects and reflects it forward to a 0.3-meter secondary mirror, which then sends the light back through a central hole in the primary mirror. Behind the primary mirror, five scientific instruments and three fine-guidance sensors, themselves complex assemblages of optics, motors, and electronics, share the bundle of precisely focused light. Surrounding the optical telescope assembly, the support systems module contains dozens of electronic and mechanical black boxes to operate the ensemble, and sprouts two deployable antennas and two large solar arrays to generate electricity.

Ensuring Hubble could be serviced in orbit was a lusterless and exacting job, a nuts-and-bolts project involving unheralded but brilliant engineers and managers like Olivier, Ron Sheffield, Michael Withey, Frank Costa, Peter Leung, and Brian Woodworth. In fact, an important early step was to figure out which *bolts* to use. (In the end, the team settled on double-height 7/16-inch bolts with hexagonal heads.) Kathy Sullivan describes the process at length in her meticulously detailed book, *Handprints on Hubble: An Astronaut's Story of Invention*. In another work, author Christopher Gainor states: "Together the astronauts and the maintenance and repair team did a top-to-bottom inspection of Hubble in its cleanroom at Lockheed in Sunnyvale, California, assessing [the telescope's] systems in terms of whether they could be repaired or replaced by astronauts wearing spacesuits. Alterations included modest ideas, such as putting labels on connectors inside HST to assist astronauts, and a major change to the Power Control Unit at the heart of HST. The unit was attached to a wall and would be nearly impossible to access during a servicing mission. With great difficulty, Sullivan, McCandless, Sheffield, and their team persuaded managers at Lockheed and Marshall [Space Center] to attach the unit to an adapter plate to make replacing the unit merely difficult."

Combining long sessions of discussion and diagramming with even longer sessions submerged in NASA's Neutral Buoyancy Simulator, practicing with their newly crafted tools, Sullivan and McCandless literally wrote the book on space telescope repair. During the project, my dad developed a tool tethering system known as the McTether that simplified the ways in which astronauts could transport and use tools in space. In order to work on the telescope—to turn a wrench, for example—an astronaut would need to be able to anchor his or her feet in order to apply torque. He and Sullivan redesigned a proposed platform with foot restraints, the Adjustable Porta-ble Foot Restraint, along with a semi-rigid tethering system for astronaut Sherpa duties, including lugging the Adjustable Portable Foot Restraint from the shuttle to a satellite work station. Versions of these pieces of hard-ware are still used by astronauts on extravehicular activities to this day.

THE HUBBLE DEPLOYMENT MISSION, which was designated STS-31 though it flew six years *after* STS-41B, was commanded by Air Force Col-onel Loren Shriver, a soft-spoken Iowan and former test pilot. Marine colonel and future NASA administrator Charlie Bolden served as pilot. Astrophysicist Steve Hawley, dubbed the "Attack Astronomer" by astro-naut Mike Mullane in his book *Riding Rockets*, joined Sullivan and my dad as a mission specialist. It was a brainy crew. Endlessly curious, Sullivan had proved herself every bit the analytical equal of her male colleagues, and having Hawley on board was said by some to be just as good as hav-ing another computer. All of the astronauts except for McCandless joined NASA in the class of 1978. The space pirates of STS-41B only six years earlier had carried themselves with a military sternness, as if they'd have been perfectly happy to *bomb* something if their primary mission didn't work out. The crew of STS-31, by contrast, was a decidedly post–Cold War group. Clad in teal polo shirts and white trousers and flashing grins at the camera for one preflight publicity photo, they looked more like grad stu-dents than Green Berets. And yet it was a serious mission; indeed, in terms of scientific significance, it may well have been the most important mission in the history of spaceflight.

Bruce McCandless wasn't leaving anything to chance this time. On STS-41B he'd suffered from what might have been Space Adaptation Syndrome, or SAS—a seasickness-like ailment, not fully understood, that seems to result from functioning in microgravity. He'd also been battling a cold, and in fact his congestion is obvious on audio recordings made at the time. Given the circumstances, it was hard to tell how much of his 1984 illness was virus and how much was SAS, so Dad prepared for both. He put himself on a diet of sports supplement powder and Wheat Thins, threw in a prophylactic shot of antihistamines, and managed to make it through the flight in good health.

STS-31 was originally scheduled to take off on April 10, 1990, but the launch was scrubbed only four minutes before liftoff due to a problem with one of the orbiter's power units. After repairs, *Discovery* left Earth on a warm, sunny morning two weeks later. My mother and sister were there to see it again, and this time I got to join them. The statistics related to the mission are fascinating. First of all, Hubble weighed 24,000 pounds. At the time, it was the largest payload a shuttle orbiter had ever hauled into space—something akin to stuffing a sofa into the back of a station wagon and then driving it around the world a few times. *Discovery* was also required to lug the orbiting observatory some 380 statute miles above Earth, making the mission the highest that an orbiter had flown. While this height was required to get Hubble clear of the planet's atmospheric distortion, and made for some arresting photography, it was also somewhat intimidating. As Sullivan points out in the documentary series *When We Left Earth*, *Discovery*'s fuel supply upon reaching its designated orbit was already more than halfway depleted.

As with any shuttle mission, there were numerous tasks to be completed and experiments to be done. The clear and consistent focus of STS-31, though, was Hubble. Scientists around the world watched expectantly as NASA attempted deployment on the first full day of the mission. It did not go well. Hawley, operating the orbiter's spindly robotic arm, had trouble plucking the satellite from *Discovery*'s payload bay and lifting it above the shuttle, where ground control could activate it. Then, once this act of positioning was achieved, one of the telescope's two solar arrays failed to fully deploy. The arrays are sets of solar cells that collect energy to power the satellite's operations. A major innovation in Hubble's design was to attach the cells to flexible sheets that could be rolled and unrolled around a central drum, like a window blind. As clever as the concept was, it didn't work as planned. The failure of the second solar array jeopardized the viability of the space telescope, which needed the energy produced by both of the arrays to keep it functioning properly in the extreme cold of space. As Mission Control watched the clock, Sullivan and my dad were directed to get ready for a spacewalk. It appeared they would need to manually unroll the errant array. They continued their prebreathing to purge their systems of nitrogen and, with Bolden's help, they donned their pressure suits. They were eagerly awaiting the call to proceed outside the orbiter. After several attempts at troubleshooting, however, engineers on the ground finally figured out that a faulty tension-monitoring module was preventing the rollout of the solar array. Once Mission Control managed to override the monitor's signal, the EVA was no longer needed. The Hubble was deployed while Sullivan and my dad were still in the airlock, waiting to go. Ironically, they never saw the awakening of the satellite they'd worked so hard to bring to life.

The balance of the STS-31 mission went smoothly. The crew shot footage with IMAX cameras, performed experiments, and posed on the middeck for the traditional midflight photo op with a dot-matrix-printed sign saying "HST Open for Business" taped to a bulkhead behind them. Return to Earth required a record-setting rocket burn, and Shriver brought the orbiter back down at Edwards Air Force Base early in the morning of April 29, 1990, easing to a stop with the aid of carbon brakes that *Discovery* was the first to use. The crew exited to a scene of sunshine and celebration.

Hubble's deployment seemed to represent a triumph for NASA. Unfortunately, it was soon discovered that the telescope's main mirror had been manufactured incorrectly and was sending back scientifically valuable but decidedly suboptimal images from space. It wasn't that the mirror was *flawed*. It was perfectly formed—but formed to slightly incorrect specifications. Once the defect was made public, the American press, never troubled by subtlety, howled with derision. Hubble quickly went from being a symbol of success to the embodiment of technological (and financial) folly. Senator Barbara Mikulski of Maryland, among others, called it a "techno-turkey." David Letterman joked about it on late-night TV; number six on his list of Top Ten Hubble Telescope Excuses was SEE IF *YOU* CAN SEE

STRAIGHT AFTER TWELVE DAYS OF DRINKING TANG. One cartoonist portrayed the satellite as a robotic Mr. Magoo, the bumbling but lovable animated character with notoriously poor eyesight. Lost in the furor over the faulty mirror was the fact that Hubble was behaving badly in other respects as well. For example, the telescope shuddered when it went from orbital night to day, a result of temperature-related effects on the solar array. A servicing mission was clearly necessary. Maintenance was one thing. McCandless, Sullivan, and a host of engineers had planned for it. The question now, though, was whether there was any way to repair such a fundamental flaw in Hubble's main mirror.

Like the pictures the satellite was sending back, Hubble's future seemed extremely unclear.

17

EYE IN THE SKY

Bruce McCandless's second mission capped his twenty-four-year career with NASA. He'd paid his dues, and, after an agonizingly long stay in spaceflight purgatory, redeemed himself and his career. He'd gone from Forgotten Astronaut to the Comeback Kid, from Buck Rogers to the Hubble Guy, and STS-41B and STS-31 are still remembered as two of the most important of the shuttle's 135 flights. And yet when he had a chance to exit the aerospace industry to do something different—buy a beer distributorship, for example, or start a software company—he balked. He could never quite get the shimmering space visions of Chesley Bonestell out of his head. Rather, after leaving NASA and resigning his commission in the Navy in August 1990, he immediately went to work as a consultant for the Space Telescope Science Institute (STSI), headquartered at Johns Hopkins University. STSI was the agency responsible for oversight of Hubble—and, now, Hubble's repair. The impetus was clear. Having worked for so many years on design and deployment of America's new orbital observatory, he wasn't about to let someone else's mistake define his legacy or obscure the wonders he and many others expected to see from the device.

As a member of STSI's Hubble Space Telescope Strategy Panel, along with a number of noted astronomers and other scientists, including Lyman Spitzer himself, his job was devising a way to repair the satellite. More specifically, Bruce McCandless's assignment involved winnowing the many repair proposals the panel received into a single practical plan, given the constraints of working in microgravity hundreds of miles from Earth in a pressure suit, helmet, and gloves, with limited time and resources. There was another, even larger constraint as well—NASA's post-*Challenger* safety-first mentality:

"There were weird ideas," McCandless says, a number of which ground control officials would never support, for safety reasons. "Could an astronaut be sent down inside the telescope to spray something on the mirror to change its curvature? There was thought about thickening the outer edges of the mirror, then re-silvering it in place. There was no lack of imagination."

As everyone knows by now, the panel's work paid off. Hubble's optics were repaired through installation of the Corrective Optics Space Telescope Axial Replacement (COSTAR) package by shuttle mission STS-61 in 1993. It was a crucial and highly scrutinized operation. NASA's reputation and perhaps even its future as an agency were on the line. Over the course of several days of space walks, astronauts Story Musgrave, Jeff Hoffman, and Kathy Thornton first replaced Hubble's balky gyroscopes. They went on to remove one of the space telescope's instruments, the high-speed photometer, and install a sort of sled with small mirrors attached to intercept light traveling through the telescope's main mirror before it reached the various other instruments. The new mirrors were manufactured so as to refocus the light to compensate for the main mirror's faulty vision; hence the widespread notion of COSTAR as a set of "contact lenses" for the satellite.

And they worked. Thanks to STS-61 and four subsequent servicing missions, the Hubble Space Telescope has functioned for more than thirty years, twice as long as its originally envisioned fifteen-year life span. It is, in fact, a far better machine now than when it started, as astronauts have over the years installed new and improved solar panels, the wide-field camera, the cosmic origins spectrograph, the space telescope imaging spectrograph, and the advanced camera for surveys. Indeed, COSTAR itself was removed from the satellite in 2009, as replacement instruments for the satellite were produced to take into account the manufacturing error in the main mirror.

As I write, Hubble continues to fly 340 miles above the Earth's surface, orbiting the planet fifteen times a day. It can "see" in three different spectra of light: visible, near infrared, and ultraviolet. It has captured stunning snapshots of galaxies previously undreamt of and allowed us new insights into the nature of the universe and how we got here—and, maybe more importantly, where we're going. Astronomers around the world can apply for observation time on the Hubble. Among the discoveries and confirmations Hubble has facilitated are the surprising, and somewhat alarming, facts that the universe is not only expanding but is expanding at increasing velocity; that the universe is around 13.8 billion years old; that black holes do exist, as astronomers and physicists had long postulated, and may actually be at the heart of most, if not all, galaxies; that Pluto has a fifth moon; and that orbiting the sun even beyond Pluto is a giant, potato-shaped asteroid that astronomers have dubbed *Arrokoth* (Powhatan for "sky"). Hubble allowed astronomers to watch Comet Shoemaker-Levy 9 plunge piecemeal into Jupiter and has detected what seems to be a massive saltwater ocean under the ice of Jupiter's moon, Ganymede. It has scanned the atmospheres of nearby exoplanets and found what star watchers believe to be the most distant galaxy ever observed, the mysterious GN-z11, located some 32 billion light years away in the constellation Ursa Major.

The so-called Deep Field study, made by means of long-exposure photography by astronomers puzzled by a seemingly empty portion of the sky, revealed galaxies gleaming in the distant darkness like mica in the sand of a creek bed—proving, said my father, that there is nowhere we can look, if we look hard enough, that is not studded with whirling stars. As journalist Ross Andersen explains:

> Much of the light caught by the Deep Field traveled to Hubble from across nearly the entire universe, from stars that burned out before the Earth had even begun to form. The oldest, most distant galaxies within the image have a chaotic, fragmentary structure, whereas the closer, brighter galaxies are symmetrical, marked with the familiar swirl of the Milky Way. From this progression, this cosmic vista, new notions about the evolution of galaxies have emerged. Gazing at the Deep Field is like mainlining the whole of time through the optic nerve, like counting by fingertip the tree rings of the universe.

Observations from Hubble have also contributed to our very limited knowledge of dark matter and dark energy, materials or perhaps simply behaviors that seem to fold around the far corners of physics into the realm of fiction.

One source reports that more than 18,000 scientific papers to date have included data from Hubble. The satellite's success has also been a sort of gateway drug for big astronomy, whetting the scientific community's appetite for further observation. The European Space Agency's Solar Orbiter was launched in February 2020, tasked with studying the stormy surface of the sun and getting the first photographs of the star's north and south poles. Perhaps the biggest project in the wings is the James Webb Space Telescope, which will soon take up a position some 940,000 miles from Earth (at one of the so-called Lagrange points between the Earth and the sun, which will allow for a relatively stationary position with respect to the two celestial bodies) and use its massive beryllium mirror to peer deeper into time and space than could have been imagined only a few decades ago. Interestingly, Webb, unlike Hubble, will not be maintainable in space, as it will simply be too far away for repair missions to be feasible. Back on Earth, the Vera C. Rubin Observatory will eventually deliver, from its perch in the highlands of north-central Chile, 500 petabytes of images from our cosmic environment. As astronomer Abraham Loeb puts it, "The experience will resemble a new subscription to a streaming media service, this one broadcasting the universe. The discoveries associated with this unprecedented flow of fresh astronomical information could shine a new light on our place in the cosmos."

ALMOST AS IMPORTANT AS the science is the beauty. Data counts, but so do diamonds. The images Hubble has sent back to Earth astonish even jaded sky searchers, much less those of us who wouldn't know dark matter from Darth Vader. We goggle at sights like the odd, hourglass-shaped Southern Crab Nebula, several thousand light years away; the shimmering columns of the Star Queen Nebula, the so-called Pillars of Creation, extended like fingers on the hand of God; and the teeming galactic petri dish of the Hubble Ultra Deep Field image, each of its points of light a million possible and impossible worlds.

Hubble's postcards are so fantastic that they sometimes seem more like a collection of fantasy art than pictures of actual phenomena. The size, shape, and sheer spectral weirdness of the images boggle the imagination and make prophets and dreamers of us all. Some of us pay therapists to tell us we're important and unique. Then we check in with Hubble so the satellite can inform us just how galactically marginal we all are. The truth is somewhere in the middle. We are small people on a tiny planet, not even situated in the center of our own local star cluster. And yet our brains can imagine the stretch and scope of at least some shadow of infinity. Because of Hubble's ability to capture images on the visible light spectrum, the development of the internet as a medium for sharing and studying photographs, and the brilliant production work of the technicians who take Hubble's data and translate it into color and depth and perspective, the satellite has become a public favorite—what astronaut John Grunsfeld called a science celebrity. When NASA indicated in 2004 that it wouldn't support any additional missions to extend Hubble's life span, reaction from around the world was scathingly negative. As astronomer Neil deGrasse Tyson put it:

> Akin to a modern version of a torch-wielding mob, there were angry editorials, op-eds, letters to the editor, and talk shows on radio and television, all urging NASA to . . . keep Hubble alive. At last, following pressure from Congress, NASA reversed its decision—Hubble would be serviced one last time.

Bruce McCandless was immensely proud of Hubble and his part in bringing it to life—and, later, in repairing it. Jim Crocker, then an employee of the Space Telescope Science Institute and later vice president of the civil space division of Lockheed Martin, was also on the panel that devised the fix for the satellite. Indeed, he is credited with figuring out the hardware that could be used to install corrective mirrors on Hubble, following an insight he had while visiting Germany and studying the faucet he was using to take a shower. He deflects some of this credit for the fix to my dad, who was, he said, the panel's nuts, bolts, screws, and wires guy. This sounds right. Bruce McCandless remembered his role in the repair of Hubble as being akin to an auto mechanic rather than a scientist. He was always confident the fix could be made. Crocker recalls him saying, "There isn't anything an astronaut can't do—only things NASA won't *let us* do." As Crocker put it in 2006: "Bruce played a key role in helping the team decide which of the numerous options presented for fixing Hubble were achievable in the space environment. Much of the credit for Hubble's restored vision, and NASA's restored credibility, goes to Bruce. In fact, *we could not have done it without him.*"

18

A NEW START

What do you do when you've done what you dreamed? My dad's ideas about life after NASA were contradictory. He wanted to leave government pay scales behind him and make some money in private industry. But he also flirted with the idea of ditching the nine-to-five life so he could just *invent* things, like Thomas Edison, Philo T. Farnsworth, and his grandfather Byron, who held several patents. He kept articles in his files like "Inventors Face Lengthy Battle For Acceptance," and "Thomas Edison and His Electric Lighting System," which carried the promising subhead, "The inventor's notebooks reveal that he was actually a skilled practitioner of systems engineering, not the tinkerer of legend." He made notes on ideas for new products, including, in 1986, "Digital Audio tape in same frequency standard (44kHz) as compact disks" and "dynamically re-legendable keys for [computer] keyboards." He had a crazy, possibly brilliant idea for how to create a rowing shell in which, through the use of gears and pulleys, rowers could use their customary propulsion technique but face forward in the boat. He explained the mechanics of it to me once, but every time I try to envision it now I feel my brain developing a small but significant cramp.

He couldn't resist suggesting product ideas to others as well. He sent a letter to the Heath Company, whose do-it-yourself Heathkits he loved. He started the letter by praising the company for its GC-1000 Most Accurate Clock kit, which he'd bought and assembled and found to be "technically superb, esthetically pleasing, and cost effective." He then noted that the Federal Communications Commission had recently deregulated subsidiary communications authorization subchannels and proposed development of a "receiver that would enable me to receive background music [through such subchannels] for my own personal relaxation while working at home." There were bound to be many more subchannels springing into existence, he said, and he for one would "certainly purchase an FM stereo receiver with SCA subcarrier capability if it were offered."

Eventually, the intellectual and financial rewards of continuing to work in the aerospace industry proved to be the stronger lure. In November 1990 he took a job as principal staff engineer with the Martin Marietta Astronautics Group in Denver. This wasn't a heck of a leap. Not coincidentally, Martin Marietta had built the Manned Maneuvering Unit and would be overseeing the planned Hubble Space Telescope repair mission. My dad had been working with engineers at the company for a quarter of a century. He was almost like family. In a move that was as close to impulsive as they ever got, my parents sold the Wimberley property and bought a cabin on a scenic parcel of pasture and pine trees in Conifer, in the Front Range of Colorado's Rocky Mountains. The place was situated high above Martin Marietta's Waterton facility on an unpaved route called High Grade Road. Bernice McCandless refused to drive on High Grade, which is scenic but terrifying. My father by contrast tore past Sampson Mountain, down through Blue Jay Gulch, and on into Deer Creek Canyon in a gigantic Dodge Ram pickup truck, reveling in the contrast between the mountain backroads and his work on missions to Mars. He'd taken to wearing leather vests. He grew a large mustache and sported bolo ties, like a barroom extra in a bad Western. Life, for the moment, was good.

In October 1994, for example, my mom wrote:

Dad and I drove over to Glenwood Springs on Saturday and participated in a Murder-Mystery evening at the Hotel Colorado. Lots

of fun—I was a flapper and Dad a gangster. It was a cocktails,
dinner, dance kind of thing and we stayed overnight... [W]e went
to the Navy–Air Force game at Colorado Springs. It was a terrible
game—Navy got beaten up. But the weather was wonderful—
the scenery beautiful and the crowd was in a good mood—what
more could you ask—especially when you're like me and don't care
who wins!

Mom made friends in the area. Even better, my sister Tracy moved from Clear Lake to Conifer and took up residence in the house next door to our parents. Mom rode her horses, hiked the mountains, and threw herself into a variety of local philanthropies. She planted a flower garden in back of the cabin and grew riotous, colorful things that thrived for the few weeks of summer the mountains received. She edited the local Kiwanis Club newsletter, the ominously titled "Shades of Shadow Mountain," and—though she denied it as a matter of principle—occasionally enjoyed herself.

Their interest in animals continued. Bruce and Bernice nursed back to health an injured fox, a broken porcupine, and several abandoned horses. In 1995 a local sheep rancher shot and killed a black bear and her two cubs on a neighbor's property. Neither the rancher nor the neighbor was in any danger, since the three bears had tried to escape human contact by scrambling up a nearby tree. Mom and Dad were disgusted by the killings—or murders, as they described them. They wrote to local authorities to urge prosecution of the rancher and attended the trial when charges were brought. There was no conceivable justification for the shootings, my father wrote, except machismo; the rancher just wanted to look like a big man in front of his kids and his neighbors. The rancher was eventually convicted on six of seven counts brought against him. My dad again took up his pen, this time to commend the prosecutor who had presented the case for Jefferson County.

WHEN MARTIN MARIETTA MERGED with the Lockheed Corporation in 1995 to form Lockheed Martin, the scope of projects my father could immerse himself in widened considerably. Lockheed was America's preeminent military aircraft contractor, owner and operator of the secretive California Skunk Works, where legendary aircraft like the P-38 Lightning, the U-2, the SR-71 Blackbird, and the F-117 Nighthawk stealth fighter were developed. While he held various titles at Lockheed Martin, Bruce McCandless apparently had a fair amount of latitude to work on whatever projects interested him. He participated in the company's tethered-satellite program, which demonstrated the feasibility of generating electrical current in space using kinetic energy. He was part of the team that purchased Russian RD-180 rocket engines after the fall of the Soviet Union, in an attempt to acquire valuable Soviet technology before others could and keep Russian scientists and technicians from going to work for rival nations such as China and Iran. When Washington attempted to block the acquisition of the engines, my dad was sent to talk to his old classmate, Senator John McCain, to explain the logic behind the move. The feds eventually relented. The innovative engines were subsequently employed, with slight modifications, as the first-stage power source for Lockheed Martin's ubiquitous and spectacularly successful Atlas V rocket. The Atlas V has been used to launch numerous military, communications, and weather satellites, a probe to Jupiter, two flights of the Cygnus vehicle to the International Space Station, and the Air Force's X-37 military space plane.

In the early nineties Bruce McCandless and one of his colleagues, planetary scientist Dr. Ben Clark, were selected to participate in a "Next Generation Access to Space" study being conducted by Johnson Space Center. Clark, a noted expert on the possibilities of extraterrestrial life, was at one time Lockheed Martin's chief scientist for space exploration systems. He conceived and developed the X-ray fluorescence spectrometers for the first geochemical analyses of Martian soil on the Viking landers. NASA asked two-person teams from each of the six leading aerospace corporations at the time to provide possible alternatives to the space shuttle. This industry/NASA working group evaluated seventeen alternative methods of replacing the shuttle. Most approaches were various winged vehicles, which were all the rage in those days. But Clark had worked with the design engineers at McDonnell-Douglas during the days of the Gemini program. He was familiar with the program's capsule design and knew that winged vehicles created all sorts of energy inefficiencies and design complications. He suggested instead that one of the seventeen architectures proposed for a shuttle replacement vehicle could include a return to ocean landings. He was initially rebuffed with the admonition that "astronauts don't do water anymore; they need a red carpet to walk on when they land." My father disagreed. He liked Clark's idea of designing a two-times scale-up of Gemini and eliminating the side hatches. Together, the men looked at the possibility of landing a capsule in Lake Okeechobee to avoid the deleterious effects of immersion in saltwater. A particularly innovative idea, Clark thought, was a way to quickly refurbish the capsule by removing the heat shield after the flight in order to get several technicians into the capsule at the same time from the bottom.

In a paper presented to the 43rd Congress of the International Astronautical Federation in 1992, McCandless and Clark proposed what they called the Reusable Ultralight Personnel Carrier (RUPC), which they described as a "reusable conical reentry vehicle and disposable adapter with propulsion and power subsystems." Their employer wasn't interested in the design at the time, because the company was focused on development of a bigger, better winged spacecraft with single-stage-to-orbit aerospike-engine technology and composite cryo-tanks (neither of which features came to fruition). And while NASA passed on the concept, it's interesting to note that the RUPC is remarkably similar to SpaceX's Dragon 2 capsule, though the RUPC was proposed nearly a decade before Elon Musk got into the space business, and twenty-eight years before the Dragon 2 made its first crewed flight to the International Space Station. McCandless and Clark were ahead of their time, as the following comparison compiled by Clark suggests:

	RUPC (1992)	DRAGON 2 (2020)
Shape	Gemini-Like	Gemini-Like
Diameter (max)	4.3	4
Sidewall Angle	18 degrees	15 degrees
Crew	6	7 (4 max for NASA)
Docking Port	End Cone	End Cone
Return to Earth	Splashdown	Splashdown
Parachutes	3	4
Displays/Controls	Monitor-based	Monitor- based

While working with NASA on the shuttle replacement project, Clark recalls that he and Bruce McCandless had to visit JSC on occasion. Once, they were going to a meeting in one of the agency's off-site buildings, and when they walked in they were met by a burly security guard sitting at a desk. My father handed over his special gold-colored badge. The guard apparently didn't realize this meant that he was talking to an astronaut. Taking in my father's bushy mustache, the guard said, "Okay, Cowboy, what are YOU here for?" It was rather derogatory, but Bruce McCandless took it in stride. He acted like he was no different than anyone in the room. That, says Clark, "is how I will always remember him. Bruce told you his technical opinion on everything, but he was kind-hearted and respectful and never manipulative in any way."

Bruce McCandless also worked with Jim Crocker and Ed Sedivy on redesign ideas for Lockheed Martin's Mars Polar Lander. The original lander and its accompanying surface penetrator probes, launched as part of the Mars Surveyor '98 program, were lost, somewhat mysteriously, in December 1999. A second version of the craft was built and stored by the company in hopes of being included on a subsequent mission. That chance came with the University of Arizona–led Phoenix Mars mission. Bruce McCandless's contribution to the effort was to question the conventional wisdom that had developed around the first machine's failure. Most observers figured the problem was related to the probe's software, which was not, as the computer folks say, particularly robust. Dad suggested instead that the problem might have resulted from the fact that the spacecraft was at the low end of its temperature range as it approached Mars. As the lander tried to separate from the orbiter to begin its descent to the surface, my father theorized, a connector failed to release because it was too cold. Crocker and Sedivy began reviewing the data and realized the old rocketeer might be on to something. They subsequently ordered appropriate design changes. On its second try, in 2008, the redesigned lander did, in fact, land successfully and began to send back important information about the Red Planet. In short, the mission was a success.

Dad fielded a proposition from a fire-suppression engineer who asked if Lockheed Martin would consider developing guided missiles to destroy burning oil wells in the Middle East, and seems to have given the idea some thought. He wrote papers on a variety of subjects. In 1997, for example, he opined on "Reliability and Crew Safety Aspects of the VentureStar Design," VentureStar being Lockheed Martin's ambitious but ultimately ill-starred next-generation reusable spacecraft, a project that was defunded in 2001. He expounded in 2000 on "Hubble Space Telescope Servicing: Design and Early Lessons Learned." He published a long essay in *The Denver Post* after the loss of the orbiter *Columbia* in the skies over Texas in February 2003—"Revere, Then Retire the Shuttle"— in which he argued that the "space truck's" outdated technology and lack of a practical escape system dictated building a new and better reusable spacecraft. (Though he didn't go so far as to advocate for any specific Lockheed Martin project, the replacement vehicle he recommended sounded a lot like the company's VentureStar, down to its environmentally clean exhaust products.) He wrote a paean to Neil Armstrong after the great man's passing in 2012, and he drafted an eloquent and somewhat unexpected defense of alleged diaper-wearing astronaut Lisa Nowak, whom he felt was being convicted of crimes in the court of public opinion before anyone had seen all the evidence.

Aside from construction of the International Space Station, the late nineties and the first two decades of the twenty-first century were dispiriting times for crewed-space-travel enthusiasts. Budgets stagnated. Programs were canceled, delayed, or tabled for further research. The *Columbia* disaster in 2003 led to another halt in shuttle flights and further erosion of morale at NASA. Changes in federal space-exploration strategies resulted, eventually, in a reliance on Russian Soyuz spacecraft to get American astronauts back and forth to the International Space Station, a development that cost the United States dearly in both dollars and prestige. Like many ex-astronauts, my father seethed at what he saw as NASA's overemphasis on safety and its lack of compelling goals. In 2009 he penned a scathing letter to Dr. John Holdren of the Office of Science and Technology Policy, stating that "we need a national goal for human spaceflight that is audacious, preemptive, and perhaps even shocking!" Dad's prescription: "Develop and use for exploration a large (e.g., 500 'short' tons mass) mid-solar system (hu) manned cruiser with a propulsion system having a specific impulse an order of magnitude higher than current chemical systems (e.g., app. 4,000 seconds) and an acceleration capacity of 0.01 'g' (i.e., greater than or equal to 0.32 ft/sec squared) or greater." Such a craft would most likely be powered, he said, by nuclear fission, and its size would need to be adequate for transportation of destination-specific "surface-access modules" to allow humans to land on destination planets.

BRUCE MCCANDLESS STARTED WORKING with the engineers at Martin Marietta in Denver in the mid-sixties. He joined the company in 1990 and formally retired in 2008, but he was still doing consulting work for Lockheed Martin in 2014, twenty-four years after his retirement from NASA. Hard as it is to fathom, my father's twenty-four-year stint at NASA was just the first half of his career in the aerospace industry. His connection with Lockheed Martin spanned almost five decades.

As the years went on, he shaved off his mustache. He still favored bolo ties, but his leather vest was replaced by a blue blazer, which he wore more or less constantly. He walked with a limp now and needed spectacles to read. He carried a quart-size plastic bag full of coins in his pocket. But his appetite for work was undiminished. When teamed with young engineers, fresh-faced kids just out of college, he beguiled them with his vast store of lore regarding the performance characteristics of military aircraft and their operators, the color of the sea off Cuba during the crisis in 1962, and the name of the man who closed the hatch on Armstrong, Aldrin, and Collins before they took off for the moon. He'd been there at the aerospace equivalent of Woodstock, and where he went he brought with him the holy incense of Aqua Velva, jet fuel, and leather flight jackets. He became something of a totem at the company. In 2018 Lockheed Martin chose to name its proposed Artemis program lunar-landing vehicle, modeled after the company's successful Phoenix and InSight Mars landing vehicles, the McCandless Lunar Lander. If the company wins a contract for delivery of payloads—scientific instruments, lightweight rovers, and so on—from a planned orbital gateway to the lunar surface, the McCandless Lunar Lander will be the vehicle delivering them. And this means, if all goes well, that a McCandless will finally get to the moon.

19

KING OF TERRORS

I n 2007 Bernice McCandless was diagnosed with breast cancer. It was
something she'd worried about for years. She seemed to have been waiting for it, as if it were a judgment she'd incurred in some previous life. My
father disliked talk of mortality. To name it, it seemed, was to summon it.
But he never flinched once the malady arrived. Bruce McCandless took the
only approach he knew: measure and countermeasure. Systems and science.
Together my parents attacked the cancer with every weapon they could
find. They managed to check the illness through chemotherapy, and my
mom lived a relatively normal life for years thereafter. In 2011, though, she
fell ill again. A PET scan found widespread osseous metastatic disease—
bone cancer.

Mom and Dad saw doctors in Denver and visited the M.D. Anderson
Cancer Center in Houston for tests and a consultation. Not long afterward, my parents invited my older daughter up to Colorado for what they
called Camp Emma—a week of activities like zip-lining (Mom passed
on this one, though Dad, at 75, joined in), paddle-boating, and a visit to
the Denver Zoo. Looking back, I realize Bernice McCandless was doing
what she could to imprint herself in Emma's mind—to make memories
before she was gone. As the cancer waxed and waned in response to various
treatments—chemo, radiation therapy, steroids—my mother grew steadily
weaker. She battled nausea and vertigo. Her skin grew thin and brittle. She
bruised easily and spent days recuperating from the latest bout of therapy.

But she did recuperate. We got used to seeing her in a turban. She laughed when she came to see us. We had hope.

In the summer of 2013 my mother was knocked over by one of her misfit animals—the crippled racehorse, it turned out—and had to be hospitalized for several days. Late one night in October, still weak, she went to the kitchen to get a pain pill. Momentarily disoriented, she fell and suffered a spiral fracture of her right femur, which had been weakened by the cancer. She never really recovered from the break. Near the end, around Thanksgiving, my dad started giving her THC for the pain, though we weren't allowed to tell her what it was, as she was a longtime marijuana foe. Worried that my mom was uncomfortable, my wife, Pati, suggested that she receive hospice care at home. I suspect my mother would have assented. But as with most decisions in their marriage, she deferred to her husband, and my dad wouldn't hear of it. At the time, once a patient went into hospice, Medicare would no longer pay for treatment—and my father wasn't willing to give up on treatment. In his view, there was no honor in surrender. *You've got to do your exercises*, he would say to my mother. *Get up. Keep moving.*

My mother had a terrible fear of dying among strangers, and always insisted she wanted to be cared for at home. So she was. It wasn't a great setup, but my dad and Tracy did everything they could. They fed and bathed and took her to physical therapy and to chemo and blood infusion appointments. My father felt guilty because the night she broke her leg, my mom had gone to get medicine that he normally set out by her bed. If he'd remembered to do it, he said, she would never have gone looking for it in the kitchen. But there was no point in that sort of speculation. Years of fighting the disease and the effects of its treatment had weakened my mom's immune system; the cancer was moving in her body like fire in a dry forest. Slowly but inexorably, death was coming—in three months or five months or seven. I guess we all knew it, but I was a thousand miles away, and I did a pretty good job of pretending it wasn't true.

As my mom's doctors deployed increasingly fearsome and sophisticated techniques to beat back the disease, her body stopped helping. My beautiful mother, bright saint of the suburbs, sponge of all anxieties, spent her days on the couch and her nights in the bedroom of my parents' cabin. By December 2013 she was calling out for her mother, Jeanette, long since deceased, and her sister Janet. Her long black slide elicited increased desperation from my father. When Pati and I visited for Christmas that year, we found my mom and dad lying together in bed in the darkness, the mountain wind howling around the eaves of the cabin. He got upset when I brought her a snack; the wrong type of food, he said, made her lips bleed. He was furious when we tried to clean the place. Any change irritated him. He wanted everything to stay the same, as if by holding the world where it was he could keep her with him. He grew increasingly frustrated with the fact that medical technology couldn't save her, and that even Bruce McCandless, the man who could fix anything, couldn't fix this. So he schemed. He read. He questioned and cajoled and harassed my mother's physicians, and in her last days he fabricated a sort of breathing mask for her, which he described to me with desperate enthusiasm and characteristic precision. He even persuaded himself that she was getting better. On January 9, 2014, he sent me an email that was, in retrospect, high-functioning fantasy:

> The orthopedic surgeon says that she can now put 50% of her body weight on her right leg. The healing appears to be going nicely. After another appointment in four weeks (and X-rays), he expects to clear her for 100% body weight on the right leg. He's O.K. with resuming chemotherapy. At the University of Colorado Hospital, Anschutz Breast (cancer) Center, [a doctor] seems very knowledgeable and helpful. He (as well as Mile High Oncology) does not think that she is strong enough to stand conventional chemotherapy; he suggests that hormone therapy may be the best way to go for now, and is further researching the situation.
>
> For some currently unknown reason, her anemia seems to be alleviating. She's gone from needing 3.5 to 4 units of blood per week to 2 units/week (i.e., one visit to the infusion center). We are hopeful that this path of improvement will continue . . .

My mom died five days later. It was the worst time any of us had lived through. The only small consolation was that she had passed away as she wished, at home and cared for, however inexpertly, by her family. She lied to me the whole time she was ill. She was feeling better, she'd say. She couldn't wait to come to Texas. I know now that she was trying to spare me from worry and aggravation, just as she had her whole life. That's understandable. Knowing my mom, it was even predictable.

BERNICE McCANDLESS WAS BORN without the gene that manufactures confidence, and at some fundamental level she didn't care about power or politics. Society was a circus to her, and not a very interesting one. Her faith in God was her strongest belief and her greatest talent was for tending to others. My mom loved her two granddaughters, Emma and Carson. She asked me about them whenever we talked. But here's the thing: She loved everyone's grandchildren. She loved the whole *idea* of grandchildren. When Pati and I visited her in Conifer the summer before she died, we realized she had as many photographs of other people's grandkids on her kitchen bulletin board as she had of her own. That was my mom—a sucker for an infant's face.

I know my parents' life together wasn't always easy. They married young, after a brief courtship, and my mother promptly began the penurious, largely thankless life of a Navy wife—and, a year afterward, mother. The golden days of Memphis and Key West gave way to the rootlessness and loneliness of Virginia Beach and college-town California. Bernice McCandless was happy for her husband when he joined the astronaut corps, but his job was also a constant source of anxiety for her. The physical dangers were bad enough; she had psychological demons to deal with as well. Was she smart enough for him? Glamorous enough? What was he doing on those nights when he was out of town and didn't call? The stories of astronaut infidelity are legion. Despite such problems, though, my parents stayed together for more than fifty years. They never gave up. And when she was gone, I saw something I never thought possible. My father, the man without fear, took off his glasses. Softly at first, and then with growing anguish, he cried.

20

THE UNTHINKABLE

My mother's death unmoored our little family. Tracy lost her best friend, life coach, and perpetual champion. My dad lost everything. Among the casualties was his conviction that life makes sense, that science is solace. He grieved for several months, spent time with my sister, puttered around his little cabin and his big garage, cluttered as it was with carburetor parts and power tools, extension cords and junction boxes. But eventually he grew restless. Just as he had with the passing of his mother and father, his response was to look away. At seventy-six, Bruce McCandless emerged from his isolation determined to make the most of his suddenly unregimented existence. A lifetime of self-discipline, strict schedules, and government paychecks, and then the slow despair of caring for my mom, had given way to financial security and a life of semi-celebrity. Though he wasn't recognizable on the street, he was well known in aerospace circles. There are still people who would rather meet Buzz Aldrin than Beyoncé, and from such souls he received a steady stream of invitations to speak, sign autographs, and sit on panels, both in the United States and abroad.

My father was lucky to find a companion to share this new life with. Ellen Shields was a fellow Coloradan and mountain dweller who had also lost a spouse to cancer; her husband, Harry, a retired federal marshal, had succumbed to a fast-moving brain cancer just a few years earlier. Ellen could relate to the frustrations involved in obtaining and paying for health

care in the American medical system, which is both technologically dazzling and agonizingly fragmented. In those first months after my mother
passed away, that was what got Ellen and my father talking. Ellen had
raised two children of her own, a son and a daughter. She was smart and
attractive and a Bible Study friend of my mom's. She knew nothing of the
American space program other than that she shared a birthday with John
Glenn. But the new relationship progressed quickly. My dad and Ellen
were married in late December 2014. I was best man at the wedding.

Bruce McCandless's wanderlust returned. He and Ellen traveled to
China and Germany, to Sweden, Scotland, and various locales in the United
States, often combining sightseeing with speaking engagements at schools
and gatherings of space enthusiasts. Seeing him in this new life triggered a
fair amount of philosophizing on my part. Growing up, I had the sense that
my dad was an unimpressive parent: absent, undemonstrative, uninterested
in what we mortals were up to, which was mostly just watching TV. Maybe
everyone has that sense about their parents at one time or another. But I've
revised my assessment in light of my own clumsy attempts at child-rearing.
Bruce McCandless was there as often as his passions and profession allowed
him to be. He would help if you asked. He walked the walk, and we were
free to follow or not.

I suspect it was difficult for him to relate to his kids' lack of focus. It
may have been difficult for him to relate to *anyone's* lack of focus. The
future electrical engineer, fighter pilot, and astronaut was drawing schematics of light bulbs and toasters at the age of two. He always had a sense
of what he wanted to do and how to do it. By the time he slowed down, in
his seventies, he'd made himself, courtesy of Hoot Gibson and Hasselblad,
an aerospace icon. He'd been to space twice. He'd learned to fly a variety
of jet aircraft and helicopters. He'd been named a distinguished graduate
of the United States Naval Academy and been given the key to Woodrow Wilson High School. He'd earned a patent for his tool-tethering
system. He'd visited forty countries, retired from the Navy as a captain,
and won a slew of commendations, including the Distinguished Service
Medal, the Legion of Merit, and the American Astronautical Society's

Victor A. Prather Award (twice). Perhaps the biggest prize was the U.S. National Aeronautics Association's Collier Trophy, the Academy Award of all things airborne, which he shared with Ed Whitsett and Bill Bollendonk for development of the MMU. Previous recipients include Glenn Curtiss, Howard Hughes, and an obscure Ohio bicycle mechanic named Orville Wright. Bruce McCandless was inducted into the U.S. Astronaut Hall of Fame in 2005.

I find myself now in the same position as many of my friends. We're all in our fifties. We're dealing with the illnesses and deaths of our parents even as we watch our kids head off to college or careers or whatever questionable activities they happen to find important at the moment. We sit at the dining room table in the quiet of an empty house, surrounded by invoices and credit card statements, and try to figure out what we've done right and wrong over the years. Inevitably, we wonder what we might have done better. We relitigate slights, real and imagined. And, if we're lucky, we remember shafts of sunlight occasionally streaming in through the windows.

My father eventually moved into Ellen's home, a beautiful house perched even higher in the Rockies, with a view to the south that seems like a song. Ellen had designed and overseen construction of the place herself. My mom and dad's little cabin on Pleasant Park Road stood empty, besieged by frequent Front Range snow storms. Dad stopped by on occasion, but he didn't stay for long. There were too many memories tugging at his sleeve, and Bruce McCandless was not a man who enjoyed consorting with his past. Besides, the roof leaked. A bear broke two windows in the bunkhouse, and my mother's garden died beneath a blanket of ice.

MY DAD AND ELLEN came to visit us in Austin for New Year's at the end of 2016. Approaching eighty, Bruce McCandless wasn't getting around so well anymore. Still, he was energetic and interested in almost everything, and he vowed to friends and family that he'd live to be a hundred. And yet there were signs over the years that he was thinking about his own mortality. In October 2013, as my mother began the last phase of her illness, my father talked to me about his distress and shock at the death of fellow astronaut Scott Carpenter. He'd taken to noting the deaths of classmates—"*d. 3/12/2012—sudden, massive heart attack*"—in the pages of his copy of the *Lucky Bag*. He finally admitted that the damage to his eardrums he had suffered as a result of years of exposure to jet engines wasn't just going to *get better*, and was fitted with a hearing aid. He did research on testosterone and clipped articles from magazines about the health benefits of eating nuts. Despite his physical fragility, his ambitions were as big as ever.

When he visited us that December, I figured we'd spend a quiet night at home. This suited me fine. I was by this point a longtime lawyer, firmly settled into a large-trousered middle age. Maybe too *well* settled. We'd drink some eggnog, I thought. We'd play Scrabble or Mexican Train and barely make it to midnight before we waddled off to bed, carrying extra calories around our waists like one of Batman's utility belts. But that wasn't the plan. It turned out Dad had bought tickets to a show on New Year's Eve. Ellen was under the weather, so she stayed home with my wife. My father and I drove to my office downtown and parked in the garage. We were running late, so we flagged down a pedicab. Dad had some trouble getting in, but he didn't want any help. He ignored the pedicab driver's outstretched hand, got himself settled, and nodded.

"Onward," he said.

The driver had dreadlocks with purple tips.

"Right on," he replied, and we went rattling through downtown Austin with Toots and the Maytals blaring from the driver's boom box. Fireworks blossomed red and green in the skies overhead. Music leaked from a dozen doorways. My old college buddy Pat Cosgrove biked down to join us, and we spent the evening drinking Lone Star and listening to Willie Nelson sing about beer halls and last calls and blue eyes cryin' in the rain. That was my father's MO. There was always something he wanted to do, somewhere he wanted to go, an ancient geological feature—Willie Nelson, for example—he wanted to see. He drove out to the middle of Nebraska in

August 2017, just four months before he died, to make sure he caught the solar eclipse in its weird and potentially mind-altering totality. He'd been to China once, but he wanted to go back. I suspect he had some ideas on how to improve the Great Wall.

The year 2017 brought a difficult summer and early fall. In August, Dad saw a doctor about his swollen prostate gland. The biopsy that followed showed that he had an aggressive form of prostate cancer. He seems to have decided almost immediately to have the gland removed. Through a contact on the board of directors of the Los Angeles County Hospital System, he reached out to a physician associated with the University of Southern California's Keck School of Medicine, a pioneer of robot-assisted prostate surgery. Physicians often counsel men my father's age against treatment of prostate cancer. As the disease generally moves slowly, conventional wisdom holds that any number of ailments are more likely to develop into problems in the time it takes for the cancer to fully manifest. But this particular cancer was more virulent than most, and my father wasn't interested in managing the illness. I suspect the experience of watching my mother die over the course of several years—slowly and then with increasing, horrible speed—persuaded him that anything would be better than passivity. He also had the sobering example of his father's long slide into paralysis to consider. So he signed on for the surgery—risky at his age—and never wavered. It was only at the end, as he was about to receive anesthesia, that he showed some inkling of doubt. He wrote down his computer password, a secret he'd never willingly shared with anyone, and handed it, without explanation, to Ellen.

The operation went well, according to the doctor. My father was released from the hospital shortly afterward but was asked, as a matter of course, to remain nearby for observation. He and Ellen were given a room in a nearby USC medical school dormitory. They bought snacks and instant coffee and settled in for a short stay, confident they would be home in Colorado for the holidays. Dad woke up early the next morning saying his elbows hurt. Ellen brought him his pain medication, along with some water, but he wanted orange juice instead. He was conversational, and she heard nothing out of the ordinary other than the odd complaint about his elbows. He raised up on his right arm to take his medicine and orange juice, but when he lay down again his eyes rolled back in his head. Ellen called 911. When she saw the lights of the ambulance, she went out on the balcony and screamed for help. Someone on the street below heard her cries and directed the paramedics to the building. Ellen didn't want to leave my father alone, so she left her phone beside him, propped the door to the apartment open, and ran to the elevator.

The paramedics worked quickly. Outside, Bruce McCandless was loaded in the back of an ambulance. Ellen rode in front, worrying as the driver asked her trivial questions to keep her distracted from the bad dream taking place behind her. As the vehicle sped through deserted streets toward the nearby Los Angeles County + USC Hospital, my father's heart stopped beating. A team of physicians at the hospital tried for forty-five minutes to revive him, but their efforts proved futile. Bruce McCandless died at the age of eighty at approximately 4:15 a.m. on December 21, 2017.

It was probably a clot that killed him. He'd been given blood thinners to minimize this possibility, and was wearing compression socks as well, but an embolism was the number-one suspect. Whatever the cause, the end came quickly. There was no protracted period of suffering, no long after-midnights lying in bed contemplating the ticking of the clock. The Redheaded Gunman came and went. Los Angeles is a big city, and there was plenty of work to do.

I PICKED UP MY phone very early that December morning and saw a voice-mail notification. I didn't recognize the area code, but I felt a vague sense of unease. The voice message got right to the point. It was a nurse calling from a hospital in Los Angeles. I heard Ellen's voice in the background. At almost the same moment, my wife walked into the kitchen. The nurse had called her number too, and Pati told me what some corner of my brain was already whispering. My father was dead.

I started making arrangements to travel to L.A. immediately. The plane left Austin at eleven that morning. After a stop in Las Vegas, I touched down at Los Angeles International Airport at 3 p.m. The next couple of days were grim and uncomfortable. After passing away at the hospital, my dad had been taken to the Los Angeles County Morgue. Ellen had been so upset and disoriented when it all happened that she wasn't even sure what hospital he'd been transported to by the EMS personnel. We figured that out, and confirmed that Bruce McCandless's body was no longer on-site. Then we located the morgue, went and identified the body, and made arrangements to have my father transported elsewhere for burial. All the while, grief and disappointment followed us around like pickpockets at a dismal carnival.

I flew back home to Texas on December 23, anxious to spend Christmas with my wife and daughters. Ellen and I left my father in the care of a funeral home in Los Angeles's Lincoln Heights neighborhood. It was a two-story wood-frame structure in the midst of a strip of grocery stores, laundromats, and Mexican and Chinese restaurants. While the facility's website displayed a photograph of an attractive blue house with a fountain and a large palm tree standing out front, the reality was less heartening. Clearly the place had seen better days—during the Second World War, maybe. The fountain was gone. The palm tree was dead. The windows on the house were taped over, apparently because painting was in progress, and the doorbell was unfastened from the doorjamb and hung on a stick attached to the eaves. We had misgivings. But the people running the place were kind and competent, and we managed to quiet our cultural panic. So it was that the last time I saw my father in Los Angeles, the world's first human satellite lay beneath a wrinkled white sheet, watched over by a painting of Jesus with his chest open wide to reveal a glowing red heart. A statue of the Virgin of Guadalupe stood to one side, and mariachi music played a little too insistently from ceiling-mounted speakers. The building appeared to have little or no security. Anyone wanting to break in and steal a dead astronaut could have done so easily. Why would someone have wanted to steal an astronaut? Okay, so that was unclear. But in my condition of grief and confusion it seemed like a legitimate threat, and I worried about it until my dad was sent off by airliner to an affiliated funeral home in Annapolis, the city that seemed, somehow, like home for him.

21

ANOTHER FUNERAL

They say Annapolis is a beautiful city, sunny in the summer and kissed by breezes off Chesapeake Bay. It's a tourist town, famous for its crab cakes and stately architecture and for its brigade of uniformed midshipmen, upholders of tradition and duty who carry their MacBooks like tiny shields.

I have no desire to go back. I attended my mother's funeral in Annapolis in January 2014; it grew gray that afternoon, and we carried her casket through a freezing rain. The memorial service for Captain Bruce McCandless took place there four years later, on January 16, 2018, in the Naval Academy Chapel. Here the iconography was familiar: not Jesus revealing a flammable heart but rather sensible flying angels looking on from above, soft-eyed and merciful but nevertheless appreciative of the many maritime victories achieved by Academy graduates over the years. My father was baptized in the Academy's famous copper-domed chapel in 1938. He attended services there for four years as a midshipman. It was there in 1960, over the crypt where John Paul Jones lies entombed in his marble-and-steel sarcophagus, that Bruce McCandless married the beatific Bernice Doyle. It was the site of my baptism a little over a year later. My mother was memorialized there before being buried in the Naval Academy cemetery plot where my father was now going to be interred. The plot was just a few feet from my grandfather Bruce's grave and the modest lane the Navy had dubbed McCandless Street several decades earlier.

Principal eulogist was Major General Charles Bolden, the retired Marine pilot and former head of NASA who'd piloted STS-31, the Hubble deployment mission. Bolden is slightly built, almost elfin, a gregarious man with a fleecy but close-cropped beard. He spoke humorously but earnestly about my dad's accomplishments and ingenuity, sneaking in a sly reference to Bruce McCandless's habit of ostentatiously clearing his throat before making an observation. He recounted the story of the balky Hasselblad camera my dad fieldstripped and put back together one afternoon, and remembered the faulty solar panel on the Hubble Space Telescope that he deduced the fix for while mid-mission on STS-31. Also in attendance were former astronauts Vance Brand and his wife, Beverly; Kathy Sullivan; Tom Jones; and Robert Cabana. Aunt Rosemary traveled up from Dallas, and my uncle Douglas came from D.C. Dr. Steve Lee flew in from Denver. My dad's Lockheed Martin boss Jim Crocker was there, as were Kathleen Willingham, the widow of Dad's Naval Academy buddy Dave, and their daughter Peg. Several members of the Class of '58 showed up to remember their fallen classmate and those long-ago days on the parade ground at Canoe U. We all hoped John McCain would make an appearance, but it wasn't to be. The senator was battling the brain cancer that would take his life just a few months later. He was in no condition to travel.

It was an overcast day with the temperature in the twenties. The sky was gun-barrel gray, and chunks of ice floated in the harbor. It had snowed the day before, and it would snow again that afternoon, but the weather held for the graveside service. Four sailors clad in black uniforms and wearing white gloves carried my father's coffin, a cherrywood casket with brass handles. The sailors fired three volleys by way of salute, and a young naval officer presented Ellen with the neatly folded American flag that had been draped over the casket. That was it. It took twenty minutes.

We adjourned afterward to the Academy's Officers' Club for a catered lunch and reception. A mild-mannered, cardigan-wearing gentleman introduced himself to me just after dessert. It was John Poindexter, who had the distinction of graduating number one, just ahead of my dad, in the Class of '58. He rose to the rank of admiral and served as President Ronald Reagan's national security advisor in the early eighties, in which position he was a bulwark of anticommunist American foreign policy and, naturally, a bête noire of the American left. As a college student, I'd criticized Poindexter and Reagan and everything they stood for. Now was the chance for me to reconnect with my younger, wilder, more virtuous self and strike a blow for the oppressed everywhere.

"I admired your father," Poindexter said.

It was the past tense that got me. I suppose I hadn't really focused yet on the fact that everything related to Bruce McCandless was now to be considered by looking back. I had the sense, somehow, that I'd just missed a late-night train. I was standing on the platform, bags in hand, but it was clear I wouldn't be going anywhere for a while. The lights of the last cars disappeared into the darkness.

"Viva Sandinista!" I wanted to shout.

But I didn't.

My dad's old classmate held out his hand, and I took it.

22

THE TRUE BELIEVER

Bruce McCandless resisted putting together an autobiography for many years. He may have thought his accomplishments weren't worth recounting. He might have felt, superstitiously, that looking back would keep him from moving forward. Whatever his reservations, he overcame them near the end of his life. He told friends and family that he would, in fact, start writing soon. He died before he had the chance, so I've done it for him, as best I could.

I hope I've conveyed a sense of the man I knew. He was an intelligent person, and intelligent people aren't easily summarized. I confess I've left out a few items. In the interest of narrative cohesion, I haven't mentioned my father's love of sushi, his interest in opera, or his ill-fated attempts to master the bagpipes. He could be moody and preoccupied; I've left that out too. He had circuits to wire, after all, and planets to see. As with all the astronauts, an appearance of ease masked a compulsive restlessness, an appetite for *motion*. Human relationships occasionally seemed like friction in this regard. But mostly he was encouraging and interested. And in the end I have to admit that, despite my best efforts, Bruce McCandless's inner life, all those gears and springs and lines of source code, remains a mystery to me.

When I was a teenager I would stand in his study, the sanctum sanctorum, and catalog his books. There was Rachel Carson's *Silent Spring*, with several pages dog-eared. Beside it stood an ornate copy of the Koran given to my great-grandfather Byron, Captain Bing Bang, when he was stationed in Turkey after World War I. There was Washington Irving's *Conquest of Granada*, present for no particular reason that I could see; Richard Henry Dana, Jr.'s *Two Years Before the Mast*, a natural choice for a Navy man; textbooks on physics and engineering; a dictionary the size of a toolbox; and an amateur radio handbook. Tucked in among these heavyweights was a slender paperback called *The April Game*, by Diogenes, a primer on how to save money on income taxes, which clued me in to the fact that my dad occasionally thought about practical matters. A gap indicated where *Open Marriage* might once have lurked, now long since exorcised from the house by my mother through incantations from *The Fannie Farmer Cookbook* and plentiful applications of White Shoulders perfume. There was a beautiful edition of *Paradise Lost*, now missing, though perhaps it will be regained, the *Handbook of Chemistry and Physics* (34th edition), and the 1958 *Lucky Bag*, the Naval Academy yearbook. I suppose I was scanning the shelves for some sort of intellectual mall map, a schematic of how and what he thought. Unfortunately, I was never able to see a system. Bruce McCandless was no more present on the shelves than a bird in last year's nest. Try as I might, the essence of the man remained hidden, as perhaps a father's true self is always hidden from a son.

Men hunger for some understanding of their fathers. One of my best friends, a man named Glenn, has a container of his dad's ashes. Every year Glenn journeys to the Davis Mountains in West Texas. There, in a secret spot, he places another handful of the remains in a cairn to mark the man's passing. My friend is an architect, well-off, esteemed in his profession. His father was an itinerant carpenter who built houses for a living, when he could keep a job. Glenn sits for a while on the rocks, and he tries to imagine his dad's life. What he thought about. Why he did the things he did, which were often irrational and disastrous. Missing work. Mouthing off. Starting love affairs and fistfights, sometimes with the same person. Though by any measure Glenn is more successful than his father, he can't help wondering if he's been true to the man's legacy. If his father, a staunch union man and antiwar activist, would have respected him if he could see Glenn now in his Patagonia vest and fancy hiking boots.

It's a common affliction, this curiosity, an ancient exercise, a yearning. In the shadowy essence of our fathers we hope we can find answers to the essential questions: *Why did you put me here? And what am I supposed to do now?* Men wonder if they're doomed to be their fathers. Men wonder if they can live up to their fathers' expectations, or redeem their fathers' failings, or simply *find* their fathers. These conflicts fill our literature and art. They inform our faith. The Bible, for example, is all about a father—*the* Father—and a long-anticipated son. Even Jesus struggles with the demands of his inheritance. Once you see the outline of this search for understanding and purpose, it shows up in documents sacred and profane. *Hamlet. The Terminator.* The Book of Exodus. In *The Aeneid*, the root document of Roman identity, Hector goes to visit his father in the underworld, seeking wisdom. In the Godfather movies, Michael Corleone wants to get as far away from his father's life as he can—until he doesn't. The Star Wars saga revolves around questions of filial loyalty and rebellion. *Game of Thrones* is essentially a story about how men and women both are defined by a father's seed, by his story, by his particular brand of madness. In his autobiography, Bruce Springsteen rages against his father's paranoia and bitterness until he comes, late in life, to understand it. And to forgive it.

The first four books of *The Odyssey* involve Telemachus's journey in search of information about his famous progenitor. When Odysseus eventually returns to Ithaca after many long years abroad, Telemachus doesn't even recognize the man. Naturally, his father has a hundred reasons for his long absence. It turns out he missed twenty of his son's birthdays because he had to conquer Troy, and then he was bewitched by not one but two sexy women with magical powers, and then monsters attacked him, and he tried some opium, and, well, it was the gods' fault, you see? It all seems a little far-fetched, but no matter. Odysseus explains that he got back as soon as he could. There's work to be done, and, like any good son, Telemachus joins in the bloody task. Present or absent, no one holds greater sway over a man than his father.

There were lots of things I admired about Bruce McCandless, but what I respected most is the way he kept moving. The way he looked forward, not back, and was eager to find the next great challenge. This was true when he was stuck in spaceflight purgatory, waiting almost two decades to leave Earth. It was true when he flew and after he retired. Even after he hit seventy-five, and his hip was hurting, and the cumulative effect of being around jet engines all his life finally ruined his hearing, he never asked for any special consideration. He never asked for anything other than insight. He just kept going.

Every boy wants to be his father. He also wants, at some point in his life, to strangle his father. In a normal life, if such a thing exists, both of these desires fade, and it becomes possible for a son to understand the man as neither a saint nor a sadist but as something in between. What makes it a little more difficult is that some men—women, too—continue to evolve. I'm not the same person I was at thirty, and my dad wasn't the same man at sixty, a committed conservationist and enthusiastic Roller-blader, that he was at twenty-five. He changed. He was never a noun. He was always a verb.

His death was a shock. Well into my middle age and wilting like a waterless houseplant, I'm still dealing with the difficulty of not having someone I can call—or, better, email—to ask questions about band saws and circuit breakers, wireless routers and electric cars. Aside from missing his competence, though, his death meant for me a final certification that even in the unlikely event I could accomplish something to rival him, he'd never know it—which, in some sense, was the point of everything I did in the first place. I still have questions. I want to know, like everyone else, how he might have reacted if the MMU had malfunctioned—if a nitrogen-gas jet had failed to shut off, for example, and he'd drifted off into space. I want to know if he ever thought about giving up. I wonder what he thought about as he drove all those miles in the decade of disco and disillusionment, sealed up in his thoughts, when we weren't allowed to listen to the radio or stick our hands out the window. Was he working out strange equations related to the fuel efficiency of the MMU? Was he remembering *his* father, dying of paralysis while still a young man, and was that why he had this incessant urge to *move*, to *do*, to *fly*? Was he thinking about anything at all other than the stripes in the road and the miles ahead? I want to know what it felt like to be him, looking out at the world.

He was one of those lucky individuals who figured out what he wanted to do early in life and was able to focus his energy and intelligence on that goal. Ideas took shape in his head like summer storms, and while they were active he liked to fold in on himself and wait for the flash of fire. This was not the time to ask for the car keys. I may have hugged him as a child, but I don't remember hugging him as an adult. Any hugs he offered were notional only—hugs in air quotes. He had enthusiasms, and some of these he shared. He had a vast and teeming intellectual and professional life, and most of this he didn't impart. Much of what I learned about his two space-flights, for example, came to me long after the fact, from my mother and from my dad's colleagues, and from the papers, photographs, and interviews he left behind. What I've gathered is that the extraordinary thing about our life together in Clear Lake was how ordinary it all was. Surely my dad, America's real-life Buck Rogers, hero of the Hubble, was awakened each morning by a robotic arm and injected with a water-soluble protein pod. Presumably he showered in a sixteen-nozzle personal hygiene module and shaved using laser-targeting technology and then hustled off to JSC via hovercraft. Why did it seem like he just climbed into his Volvo and backed down the driveway? Sometimes I look at snapshots and film footage of him as if I were seeing him for the first time. I think, *that's* the guy who left the shuttle to fly off, untethered, into space? I used to beat him at Scrabble.

Hoot Gibson's great photograph of Bruce McCandless adorns an astonishing variety of objects: beer cozies and coffee mugs, posters and panel trucks, key fobs and yoga pants. True, Dad's Bonestellian vision of platoons of engineers outfitted with jetpacks assembling space stations in Earth orbit hasn't come to pass. The MMU seems at present to be more of a glittery technological detour than a true leap forward, but SAFER remains a useful reminder of his work, and space exploration is just beginning. I suspect the MMU, or its next iteration, will eventually be used again. And as for his other major contribution to space science, it's hard to overstate the importance of the Hubble Space Telescope. Bruce McCandless did as much as anyone to get that remarkable hardware deployed and, later, repaired. The world continues to enjoy the benefits of that work.

IN DOCUMENTARIES ABOUT AMERICAN spaceflight, astronauts frequently appear on camera to reminisce. Despite their advanced ages, they are a good-looking group of people, and they show up in attractive blazers and natty golf shirts. Bruce McCandless by contrast often appears in his NASA-issue blue flight jacket. He wore a Hubble Space Telescope patch on that jacket until the day he died, and he never ceased to marvel at the visions the satellite sent back to Earth from its perch in the sky. And yet, even when reflecting on these accomplishments, Bruce McCandless grew restless. He spoke of how he envied young people who were born at the right time to dream about flying to Mars. He wanted the International Space Station to continue its work. He wanted Americans to return to the moon to explore its highlands, to walk on the far side of our satellite and sample the water we now know to be there. He noted that the moon might well harbor valuable deposits of the Helium 3 that could be used in nuclear fusion reactors to propel us ever further from our home planet. He wanted us to visit asteroids. He knew space exploration wasn't over just because he was done with it. In fact, our journey to the stars is just beginning.

Bruce McCandless was a transitional figure in the space program. When he joined NASA, the astronaut corps was all male, dominated by fighter jocks and test pilots, a white-man's club fueled by testosterone cocktails. Halfway through his twenty-four-year career there, NASA threw open the clubhouse doors. The thirty-five individuals selected to be astronauts in Group 8 in 1978 included women, African Americans, an Asian American, and two Jewish Americans—in short, all manner of Americans. It was the start of a trend. By 1990, when my dad retired, the astronaut corps was impressively diverse. I suspect my dad was more at home with this later group of space travelers than he was with the former. Though he'd been a carrier pilot, and loved to fly, he was never really the fighter-jock type. A Volvo-driving, birdhouse-building camera bug who liked to assemble home computers, he was more comfortable with the scientists and doctors who came to the program later. He was more Spock than Kirk; more Holmes than Watson. True, he resembled Jack Aubrey more than Stephen Maturin, but Aubrey had nerdish enthusiasms as well—the trim of a fine frigate's sails, his craft's weight and feel and responsiveness—analogous to the sorts of things my dad enthused about when he described his beloved space shuttle.

The sixties and early seventies were a curious time. It's hard to think of an analogy to that era's lionization of the astronauts by the American government and an enthusiastically compliant press. They were official heroes. It was Line One of their job description. They were star warriors in crew cuts and Ban-Lon, and their exploits—John Glenn's first orbit, Ed White's space walk—were celebrated as individual feats of daring and skill at least as much as they were considered collective accomplishments. Bruce McCandless got a taste of that sort of acclaim. Thanks to a lot of hard work, a little luck, and some elegant camera work by a crewmate, he's remembered as the guy in the jetpack—the first man to fly, astonishingly and beautifully, untethered in space. And yet his bigger contribution to the space program in the long run was his part in a mission that he received almost no recognition for, the design and deployment of Hubble, and his follow-up work on fixing that overachieving satellite. The lack of attention would have been fine with him. My dad enjoyed the brief fame he earned from his jetpack maneuvers, but I think he enjoyed Hubble's enduring success even more.

Aside from occasional firsts—Scott Kelly's yearlong sojourn in low-earth orbit, for example, or Chris Hadfield's acoustic Bowie cover in space—the exploits of American and other astronauts are mostly anonymous these days. We see our astronauts preparing for launch, or floating around in the microgravity of the International Space Station, and they look not like demigods, but like neighbors and cousins and friends. That may be the point. Today's American astronauts aren't really like us. They're enormously intelligent and talented, and they're fanatically dedicated to their jobs. Still, they look like us, and they talk like us, and that's enough to get anyone, male or female, Anglo or Asian or African American, dreaming about what they can accomplish in space. This seems right to me. Not every hero has a crew cut.

WRITING ABOUT MY MOTHER is a joy. I carry her memory with me like a holy relic, and I only wish I could better re-create her warmth and illogic and bouts of self-recrimination. My dad is harder. Even at the age of eighty, Bruce McCandless was independent and headstrong. After he died, Ellen, Tracy, and I knew what we were supposed to do. We had to get him to the Naval Academy to lie in the ground beside my mother and my grandfather Bruce. There he could rest near where his grandfathers marched a hundred years before. Where he earned his engineering degree. Where he roughhoused with Dave Willingham and the guys in the Radio Club and heard the faint staccato song of Sputnik calling him like a siren on a distant sea.

He followed that call for the rest of his life. In doing so, he made himself a symbol for thousands of people who hear the same thing: engineers and astronauts, students and scientists and some kid in Sierra Leone who wonders what it's *like* out there, a little closer to the stars. My memories aren't as clear as they used to be, but some images of the man stand out still. I remember Bruce McCandless tossing a decorative pillow my way when I'm five years old, grinning as he destroys my army of plastic infantrymen. He's sprinting after a Continental Trailways bus with my camp money on a steamy summer morning in downtown Houston. He's standing with a Jersey girl named Bernie on a beach in Washington State and they've grown old together and they've paused there with their backs to the Pacific

Ocean, temporarily halted because, for the moment, there's no way to go any farther.

And yes, I see him the same way you do. It's impossible not to. He's looking back at us, a few feet away from *Challenger*. At least we assume he's looking at us: The gold solar shield attached to his helmet is down, and it's impossible to see his eyes. The machine he's wearing works beautifully. He starts moving away, still facing the spacecraft, and he slowly gets smaller, a snow white image against the blackness of space. At a hundred meters he fires the nitrogen-gas jets on the back of the maneuvering unit and eases to a stop, relative to the orbiter. And there, traveling at twenty-three times the speed of sound, suspended in the cosmos, he waits, as if to say *Join me*. It's a vision of hope and wonder. The ant and the ocean. A seed of the divine in the desert of the infinite.

But I see something else in the picture as well. I see the man who named me. He's shut off from me now, mute and unattainable, sealed up in his pressure suit, as much a mystery to me in this vision as he ever was. As much a mystery as any man is to his son, who spends his life reading the clues a father left behind and remembering his words as he tries, a hundred times, to invent his own life. I don't remember all those words, but I do hear one.

It resonates to this day.

Onward.

NOTES

1. THE LONG DRIVE

The writer called Bruce McCandless The article quoted is "Skylab Communicator Is Forgotten Astronaut," uncredited, published on page 10 of the *Rocky Mountain News* on November 23, 1973—the day after Thanksgiving. As this was an Associated Press wire story, it was picked up by a number of other newspapers, including the *Los Angeles Times* ("Forgotten Astronauts: They Still Wait for Chance to Fly in Space"), *Washington Star-News* ("Rookie Still on Earth"), and *Santa Cruz Sentinel* ("The Rookie Who Never Got a Mission"), and published under different titles, sometimes with a photograph of a pensive Bruce McCandless sitting at a mission-control communications console. Though my dad was dismissive of the press and not inclined to acknowledge criticism, he kept five copies of the article in his files, possibly for motivational purposes. It wasn't that he went out and found these copies; rather, they were mailed to him by well-meaning friends and colleagues. Such friends would also send chatty letters including language like the following: "After reading and seeing everything that Apollo XV has accomplished, we're wondering, when's your turn, Bruce?" When indeed? The article was written by William Stockton.

2. BATHED IN BLOOD

Captain Bing Bang One source says instead that Byron McCandless earned the moniker during World War I, when he toured the Mare Island Shipyard in California. The keel for the USS *Caldwell*, a destroyer he had been given command of, had recently been laid. Byron was so impatient to take command of the vessel that he joined in the construction effort. Josephus Daniels, *Our Navy at War* (New York: George H. Doran Company, 1922), pp. 66–67. Either way, the nickname reflects Byron McCandless's penchant for tinkering and manual labor.

"the biggest manmade object ever created" James D. Hornfischer, *Neptune's Inferno: The U.S. Navy at Guadalcanal* (New York: Random House, 2011), p. 276. Hornfischer's book about the war in the Solomon Islands is one of the best. It contains, on pages 282–283, this memorable description of the First Naval Battle of Guadalcanal: "On that night, two groups of powerful steel machines surprised each other on the sea in the dark and, blundering and veering in a manner unworthy of the elegance of their design, grappled bodily, delivering hammer blows until death."

Roosevelt called the engagement "Heroic Cruiser S.F. Here!" *San Francisco Call-Bulletin*, December 11, 1942, p. 2.

Nothing in the long and hallowed history Robert Brumby, "Officer Tells of Saving U.S. Fleet," *San Francisco Examiner*, December 4, 1942, p. 9.

HIS DAD'S REAL HERO *Long Beach Press-Telegram*, December 12, 1942, p. 1.

Though two of the planes For an eyewitness account of the April 8, 1945, kamikaze attacks on the USS *Gregory*, see excerpt from The Diary of William E. Bletso published on the website of the USS *Gregory* Association, https://www.USSGregory.com.

Cmdr. McCandless Takes Ship to Father's Yard *Washington Evening Star*, June 18, 1945, p. A-10. Byron McCandless's quote is reported in the article.

3. THE BOY WITH TWO BRAINS

"always confident and ambitious" Daniel Winkel, "Wilson High Enjoying Its Worldly Alumnus," *Long Beach Press-Telegram*, February 8, 1984, p. B4.

It was a brilliant moment David Halberstam, *The Fifties* (New York: Villard, 1993), p. 487.

a "sturdy conversationalist and party man" is a characterization of McCain by the 1958 edition of *Lucky Bag*, the Naval Academy yearbook.

Reef Points is still distributed to all incoming midshipmen, who are expected to memorize its contents. John McCain and Frank Gamboa, each of whom were members of the Class of '58, both spend time discussing the Academy's educational and disciplinary rituals in their books, *Faith of My Fathers: A Family Memoir* and *El Capitan! The Making of an American Naval Officer*, respectively. McCain was almost expelled for disciplinary reasons. Gamboa comments that he himself "spent many a Saturday afternoon of my plebe, youngster, and segundo years marching on the terrace" as punishment for various administrative infractions. Frank Gamboa, *El Capitan! The Making of an American Naval Officer* (Jacksonville, Florida: Fortis Publishing, 2011), p. 66.

the astonishing feat of traveling *under* the North Pole The *Nautilus*'s subpolar cruise was itself a response to the Soviets' new space surveillance threat. Even the most powerful satellite cameras couldn't see underwater, after all, and the *Nautilus* could stay submerged for weeks at a time. It could also, one Navy observer wrote with regard to the vessel's 1957 sortie under the ice, surface and still be camouflaged by arctic pack ice and thus remain virtually undetectable. Cdr. Robert D. McWethy, USN, "Significance of the Nautilus's Polar Cruise," *Proceedings*, May 1958, Vol. 84/5/663.

4. YOU ARE ON FIRE

Ike, balding and avuncular https://www.presidency.ucsb.edu/address-us-naval -academy-commencement.

It didn't take long for Bruce McCandless The information about Bruce McCandless's Skyray mishap comes from a copy of the U.S. Navy's official investigative file regarding the incident.

Lt. Tom Anderson "VF-102 Pilot Rescued by Helicopter Crew," *Forrestal Antenna*, February 21, 1961, p. 2.

The confrontation occurred when the Soviet Union Benjamin Schwartz, "The Real Cuban Missile Crisis," *The Atlantic*, January/February 2013, https://www.theatlantic .com/magazine/archive/2013/01/the-real-cuban-missile-crisis/309190/.

Richard Nixon had just lost his race "Beaten Nixon Accuses Press," *The Virginian-Pilot*, November 8, 1962, p. 1.

3,039 miles I know the mileage because my dad kept his reimbursement voucher for the trip, as he kept almost every piece of paper he ever touched. He received reimbursement from the Navy at the rate of six cents per mile.

5. DON'T TREAT HIM AS A PILOT

Lantern-jawed Mississippian Fred Haise Michael Collins, *Carrying the Fire: An Astronaut's Journey* (New York: Farrar, Straus and Giroux, 1974), p. 180.

SPACE TEAM GETS 19 "Space Team Gets 19 New Astronauts," *Orlando Sentinel*, April 5, 1966, p. 3-A. "I'm Off to Moon, Dear," *Oakland Tribune*, April 5, 1966, p. E-3. "New Astronaut Got Space Bug When He Was 10," *Chico Enterprise-Record*, April 5, 1966, p. 5.

Other observers were more biting Rick Houston and Milt Heflin, *Go, Flight! The Unsung Heroes of Mission Control, 1965–1992* (Lincoln: University of Nebraska Press, 2015), p. 13. The disparaging adjective was also assigned to the following astronaut group, Group 6, which became widely known as the Excess 11.

effectively treated as a "scientist-astronauts" Matthew Hersch, *Inventing the American Astronaut* (New York: Palgrave MacMillan, 2012), p. 86.

heat, humidity, smog Brian O'Leary, "Rebellion Among the Astronauts," *Ladies' Home Journal*, March 1970, p. 143. In the same article, O'Leary recounted telling Deke Slayton that flying just wasn't his "cup of tea"; indicated that he was impatient with the sight of Apollo astronauts "hopping around" on the moon; and recalled his joy at visiting San Francisco on a "lazy day," where "the hippies were a beautiful sight, peacefully walking, rolling, sitting or lying in the grass under a brilliant warm sun." NASA might as well have invited George Harrison to join the space program.

fifty clean-cut, erect, alert *Ibid.*, p. 143.

a metallic star pasted on his forehead "Ex-Astronaut Warns as Dancers Chant," *Los Angeles Times*, April 15, 1979, p. 38.

dismayed that his star pupil Andre Neu, "Mountain View Resident Selected as Astronaut," *Standard Register Leader*, April 5, 1966, p. 1.

6. YOU'LL GET USED TO THE HEAT

So Clear Lake got the nod Not everyone was happy about the selection. When the first official NASA delegation arrived in Houston in September 1961, they wondered what they'd gotten themselves into. Carla's wreckage still littered old Highway 3, and "the night was humid . . . [the] air . . . heavy with odor of the Houston channel, [and] industries blowing downwind from petroleum and chemical facilities and the paper mill in that general area." Henry C. Dethloff, *Suddenly, Tomorrow Came: The NASA History of the Johnson Space Center* (Mineola, New York: Dover Publications, Inc., 2012), p. 41.

He and his colleagues rafted ASTRONAUTS RIDE THE RAPIDS, *The Houston Post*, June 20, 1967, p. 20. Interestingly, the Chagres is the only river in the world that drains into both the Atlantic and Pacific Oceans. Judging by the grins on the faces of the astronauts, the ride wasn't particularly strenuous.

space-related expenditures spiked Arnold S. Levine, "Chapter 4: The NASA Acquisition Process: Contracting For Research and Development," *Managing NASA in the Apollo Era: An Administrative History of the U.S. Civilian Space Program*, 1958–1969, Scientific and Technical Information Branch, National Aeronautics and Space Administration.

The change was mightier than he had counted on Norman Mailer, *Of a Fire on the Moon* (New York: Random House, 2014), p. 54.

fiendishly difficult to operate Tom Jones, "Bruce McCandless and His Flying Machine," *Air & Space Magazine*, February 23, 2018, https://www.airspacemag.com/daily -planet/bruce-mccandless-and-his-flying-machine-180968241.

Mike Collins called it Michael Collins, *Carrying the Fire*, p. 127. Even after forty-seven years, Collins's book is the gold standard of astronaut autobiographies, funny, compelling, and endearingly geeky. Consider his description of the lunar module, which the crew named *Eagle*: "It doesn't look like any eagle I have ever seen. It is the weirdest-looking contraption ever to invade the sky, floating there with its legs awkwardly jutting out above a body which has neither symmetry nor grace. Everything seems to be stuck on at the wrong angle, which I suppose is what happens when you turn aeronautical engineers loose designing a vehicle which always flies in a vacuum and hence requires no streamlining." *Ibid.*, p. 398.

The AMU had twelve thruster nozzles David C. Schultz, "Astronaut Maneuvering Unit Operational Summary," NASA Internal Document, December 21, 1965, and W. C. McMillin and Maj. E. G. Givens, Jr., "Description and Status of DOD Gemini Experiment D-12 Astronaut Maneuvering Unit (AMU)," NASA Internal Document, Undated. Ed Givens, another Group 5 astronaut, was originally given primary responsibility for development of the AMU. My father's involvement increased after Givens's death in a car accident in June 1967.

The heat generated by the H_2O_2 Eugene Cernan and Don Davis, *The Last Man on the Moon* (New York: St. Martin's Press, 1999), p. 133.

His description of the EVA Cernan's failed attempt to test the AMU on Gemini 9 was an embarrassment to him, and he spends a fair amount of time in his autobiography discussing what happened and why. Cernan and Davis, *The Last Man on the Moon*, pp. 133–135. He was hindered in part by perception. Ed White's spacewalk on Gemini 4 had seemed to be effortless and enjoyable. Cernan, on his "space walk from hell," as he called it, had much more to do and no handheld propulsion unit, as White had, to help him do it. Despite the differences, NASA officials, including Gene Kranz, had trouble understanding the discrepancy in the two astronauts' performances, and developed what Kranz called a "decided and long-lasting lack of enthusiasm for EVAs." Gene Kranz, *Failure Is Not an Option: Mission Control from Mercury to Apollo 13 and Beyond* (New York: Simon & Schuster, 2000), p. 188. This dislike continued well into the 1980s. A NASA memorandum from 1988 sounds the refrain. Advising engineers working on plans for Space Station Freedom, the memo states that EVAs are "time-consuming and labor-intensive," "more difficult," and pose "an additional crew risk." Thus, "in no event should EVA be encouraged unless it is clearly required and in the best interests of the program." The distrust may have been even more intense for *untethered* EVAs. As one newspaper noted shortly after Cernan's aborted jetpack test, "abbreviation of . . . Cernan's Sunday morning walk in space virtually eliminated the possibility a future Gemini pilot will be unleashed in orbit." Rudy Abrahamson, "Shortened Walk Casts Doubt on Untethered Man in Space," *Los Angeles Times*, June 6, 1966, p. 1.

A photograph of him in *The Sacramento Bee*, December 15, 1968, p. 18.

Operating machinery while in the clutches Neutral buoyancy testing is the best technique for simulating extravehicular activity in microgravity. Astronauts train underwater in an oxygen-pressurized suit. The oxygen makes the suit, and thus the astronaut, float. The astronaut is then ballasted with weights that balance out the buoyancy and allow the astronaut to remain at a given depth indefinitely. The method is not completely true to spaceflight, of course. Because the pool is on Earth, gravity applies, so that an astronaut who turns upside down in the pool will have the blood rush to his or her head, an uncomfortable and generally unsustainable phenomenon. The original neutral buoyancy facility was located at Marshall Space Center in Huntsville, Alabama. It was decommissioned in 1997 and is now a National Historic Landmark. JSC had a smaller neutral buoyancy tank, called the Weightless Environment Training Facility, until it was replaced by the Neutral Buoyancy Laboratory, which the astronauts currently use.

Thus, the biggest plum of his early career All told, Bruce McCandless did capcom duty for Apollo missions 10, 11, and 14, and all three crewed Skylab missions.

For today, we'd like you on Transcript and audio recording of Apollo 11 Mission Control communications, available on the wonderful Apollo in Real Time website, https:www/apolloinrealtime.org/11/.

After the Eagle touched down In Jim Hansen's great biography of the man, Neil Armstrong suggested that the plan all along was to proceed with the space walk without a rest period and that the built-in siesta was really just a buffer in case the astronauts got behind on other tasks. If such was the case, it was a surprise to Bruce McCandless. James R. Hansen, *First Man: The Life of Neil A. Armstrong* (New York: Simon & Schuster, 2012), p. 486.

mundane and magnificent Transcript of Apollo 11 Mission Control communications, available at https://www.apolloinrealtime.org/11/. Mike Collins describes a moment that took place shortly afterward: "Houston comes on the air, not the least bit ruffled, and announces that the President of the United States would like to talk to Neil and Buzz. 'That would be an honor,' says Neil, with characteristic dignity. 'Go ahead, Mr. President. This is Houston Out,' says Bruce McCandless, the Capcom, as if he instructed Presidents every day. The only clue as to how he must feel comes in his use of the word 'out,' which we are all taught in telecommunications protocol but which we practically never use. 'Out' has a formality and finality that renders its use most unusual. Perhaps its use should be reserved for Presidents." Collins, *Carrying the Fire*, pp. 408–409.

I think it is equal in importance Quoted in Miles O'Brien, "We Aimed for the Stars . . . Until We Stopped," *Space News*, January 22, 2009, https://spacenews.com/oped-we-aimed-starsuntil-we-stopped/.

For one fleeting moment Julie Wittes Schlack, "The Fleeting Hope and Wonder of Man on the Moon," https://www.wbur.org/cognoscenti/2019/07/18/50th-anniversary-moon-landing-julie-wittes-schlack.

the greatest week in the history of the world Quoted on the Smithsonian Institute's Air and Space Museum website, in "Apollo 11 and the World," July 15, 2009, https://airandspace.si.edu/stories/editorial/apollo-11-and-world.

7. TRY STAYING HOME

With the exception of his sister Rosemary, though Bruce McCandless wasn't always focused on family. Nevertheless, after his parents died, he tried to play big brother to his siblings. My aunt Rosemary remembers, "When I graduated from Colorado College in 1971, I went to the graduation in Colorado Springs. Nobody else attended. My stepfather and his third wife were not interested, and my husband and his family didn't think it important to go. As I was getting ready to go up on the stage to get my diploma, your dad appeared! He had borrowed an astronaut jet and flown up that morning. I will never forget how much that meant to me."

8. VOLUNTARY COMPLACENCE

Though he was more social than my mom While my parents generally kept to themselves, they were nevertheless charter members of a group of neighbors—the Smiths, Morrises, Geehans, Kurzes, Ganters, Minars, Sienkowskis, Taylors, Nakamuras, and Bakoses—that called itself the Lower Whitecap Recreation Association, after the name of the street (Whitecap Drive) we lived on. Young, gregarious immigrants to Texas from various states, the men and women of the association gathered for cookouts, holiday parties, and Apollo splashdown observances in the late sixties and early seventies. Sometimes my dad showed 16-millimeter NASA movies on a screen in our living room. Kids sprawled out on our avocado-colored carpet to marvel at the flame trails generated by the Titan and Redstone rockets. "Backwards!" we'd shout. "Make it go backwards!" These days I think of this cry every time I see a Falcon 9 rocket landing. Other times the grownups met to drink Cuba libres and listen to Herb Alpert and the Tijuana Brass LPs at the Smiths' house. Membership in the association was voluntary but difficult to resist. As funding for the space program dried up, though, the neighborhood changed. Many of our neighbors moved away in the early seventies in search of better prospects. My parents never really replaced those friendships, and their social lives suffered.

9. APPALACHIA UP ABOVE

ISN'T LIKE IT ONCE WAS David T. Langford, "Man's Leap for Moon Isn't Like It Once Was," *Miami Herald*, January 29, 1971, p. 1-B.

he wasn't even given a slot on an *imaginary* flight Donald K. "Deke" Slayton and Michael Cassutt, *Deke! U.S. Manned Space: From Mercury to the Shuttle* (New York: A Tom Doherty Associates Book, 1999), p. 339.

Each probe carries a golden phonograph According to NASA's Jet Propulsion Laboratory website, "The contents of the record were selected for NASA by a committee chaired by Carl Sagan of Cornell University and others. Dr. Sagan and his associates assembled 115 images and a variety of natural sounds, such as those made by surf, wind and thunder, birds, whales, and other animals. To this they added musical selections from different cultures and eras, and spoken greetings from Earth people in fifty-five languages, and printed messages from President Carter and U.N. Secretary General Waldheim. Each record is encased in a protective aluminum jacket, together with a cartridge and a needle. Instructions, in symbolic language, explain the origin of the spacecraft and indicate how the record is to be played. The 115 images are encoded in analog form." https://voyager.jpl.nasa.gov/golden-record/whats-on-the-record/.

"Skylab Crew Takes Over Space Endurance Record," *The Greenville News*, June 19, 1973, p. 1.

different from flying a spacecraft Craig Covault, "Skylab Aids Design of Maneuvering Unit," *Aviation Week & Space Technology*, June 3, 1974, p. 45.

"Not only did I have difficulty" Bob Andrepont, Skylab 1/3 Voice Transcription, Part 2 of 4, p. 144, https://www.scribd.com/document/49132513/Skylab-1-3-Onboard-Voice-Transcription-Part-2-of-4.

The rate-gyro system Covault, "Skylab Aids Design of Maneuvering Unit," p. 45.

The ASMU was flown in various modes by all three astronauts There is some evidence that Paul Weitz experimented with the jetpack on SL-2 as well, though he wasn't supposed to. "Skylab Crew Takes Over Space Endurance Record," *The Greenville News*, June 19, 1973, p. 1.

"I would be willing to take [it] outside" Andrepont, Skylab 1/3 Voice Transcription, Part 2 of 4, p. 146, https://www.scribd.com/document/49132513/Skylab-1-3-Onboard -Voice-Transcription-Part-2-of-4.

The more than 40,000 photographs Leland F. Belew, Ed., *Skylab: Our First Space Station*, Ch. 10, "The Legacy of Skylab," 1977, https://history.nasa.gov/SP-400/ch10.htm.

forbidden from calling the subjects of their photographs "targets" Kevin Rusnak, NASA Johnson Space Center Oral History Project, Interview of Dr. Joseph P. Kerwin, May 12, 2000, https://historycollection.jsc.nasa.gov/JSCHistoryPortal/history /oral_histories/KerwinJP/KerwinJP_5-12-00.htm.

The heat-trapping nature of carbon dioxide Global Climate Change Fact Sheet, https://climate.nasa.gov/evidence, retrieved September 27, 2020.

As Joe Allen comments Joseph P. Allen, *Entering Space: An Astronaut's Odyssey* (New York: Stewart, Tabori & Chang, 1984), p. 84.

11. THE FORGOTTEN ASTRONAUT

I got to working on experiments Charles D. Benson and W. D. Compton, Skylab Interview with Cdr. Bruce McCandless, August 5, 1975, p. 4. NASA Historical Office Publication.

Why won't Bruce McCandless Walter Cunningham, *The All-American Boys: An Insider's Look at the U.S. Space Program* (New York: ipicturebooks, 2nd edition, 2010), p. 47.

"a very unusual character" Jennifer Ross-Nazzal, NASA Johnson Space Center Oral History Project, Interview with James D. A. "Ox" van Hoften, December 5, 2007, pp. 24–25.

"a little difficult, sometimes" Jennifer Ross-Nazzal, NASA Johnson Space Center Oral History Project, Interview with Robert L. "Hoot" Gibson, January 22, 2016, p. 23.

Apollo 12 moonwalker Al Bean Alan Bean quoted in Thomas O'Toole, "Astronaut Sold NASA on Jet-Pack Flight," *The Washington Post*, February 7, 1984, p. 2A. O'Toole also reports in the article that McCandless liked to go backpacking alone

in New Mexico and Arizona, that some said he seemed "out of step" with his comrades, and that "other astronauts said that he seemed to be more interested in the
fate of endangered bird species than in the course of manned space flight."

The reasoning behind the particular technical assignment Kathryn D. Sullivan, *Handprints on Hubble: An Astronaut's Story of Invention* (Cambridge: The MIT Press,
2019), p. 32.

Rick Houston and Milt Heflin suggest Houston and Heflin, *Go, Flight! The Unsung
Heroes of Mission Control, 1965–1992*, p. 174. Armstrong's heartbeat evidently rose
to 160 beats per minute, up from a resting rate of 60, toward the end of his time
on his moon but before his last scheduled rock-collecting assignment. This may
have been when Slayton advised McCandless to terminate the EVA. *See* notes at
111:32:02 of elapsed mission time in NASA's transcript of the Mission Control's
communications with the mission, https://history.nasa.gov/alsj/a11/a11.clsout.
html. As Houston and Heflin point out, Slayton's interference put the young astronaut in a tight spot. Should he defer to the boss, or follow the EVA checklist, as
planned? "I'd only been in the program three years," McCandless later said, "so
[Deke] was like God, almost."

12. THE LOON

those famous Earthrise photographs This quote and the *Last Whole Earth Catalog*
quote, Matthew Hersch, *Inventing the American Astronaut*, p. 136.

I suspect he felt the same way about words It's my belief that Bruce McCandless's narrative probity held him back professionally. He was a fascinating man, unflinchingly
honest, with an interest in everything. He had beautiful stories to tell. But he had
trouble communicating his passions, which tended to sound more like measurements. Advocacy is in some respects the art of exaggeration. He was incapable of such
embroidery, which made him a valuable resource but a less than compelling speaker.

This leads me to the concept of Spaceship Earth "Meeting Captain Bruce McCandless II," The Rogue Astronaut Blog, https://therogueastronaut.com/2017/11/13/
meeting-capt-bruce-mccandless-ii/, November 13, 2017. The Spaceship Earth
concept was popular in the seventies. It has a number of antecedents, but one

important source was the mathematician and philosopher Buckminster Fuller, who wrote a book in 1968 titled *Operating Manual for Spaceship Earth*. Dad's vision of the planet as a sort of machine, with systems and valves, aligns closely with Fuller's.

concerned citizenship more than anything Harold Scarlett, "The Environmentalists," *The Magazine of The Houston Post*, January 27, 1985, p. 9.

13. BUCK ROGERS AND THE SILVER BIRD

In the 1920s a comic-strip character Dick Calkin and Phil Nowlan, *The Collected Works of Buck Rogers in the 25th Century* (New York: Bonanza Books, 1969), p. 2.

"McCandless," said the article William Stockton, "Rookie Still on Earth," *Washington Star-News*, November 23, 1973, p. A-2.

eleven, by one count Thomas O'Toole, "Astronaut Sold NASA on Jet-Pack Mission," *The Washington Post*, February 7, 1984, p. 2A. O'Toole quotes Ed Whitsett in the article as follows: "When this mission was proposed a year ago as a rehearsal for using the backpack to repair a damaged satellite in space, engineers pointed out that the backpack, as then constructed, did not have the thrusting power to stabilize a damaged satellite. Bruce suggested we change the thruster computer logic to stabilize the satellite . . . Not only did he do it, but he came up with such a simple way to do it that it cost us almost nothing. All I can say about Bruce McCandless is that he is a thorough, methodical and brilliant electronics genius."

pistol-grip hand controllers Andrew Chaikin, "Untethered," *Air & Space* Magazine, October 2014, https://www.airspacemag.com/space/untethered-180952792/.

a brilliant engineer Kathryn D. Sullivan, *Handprints on Hubble*, p. 102.

The MMU got a big boost in 1979 Ben Evans, "To Face Their Wives, Part 1," *AmericaSpace*, February 1, 2014, https://www.americaspace.com/2014/02/01/to-face-their-wives-30-years-since-the-first-untethered-spacewalk-part-1/.

the same year it awarded the company a $2 million contract William Harwood, "1980 NASA Contract Awarded for Tile Repair Kit," *Spaceflight Now*, February 21, 2003, https://spaceflightnow.com/shuttle/sts107/030221tpsrepair/.

In the meantime, though, the orbiter was completely retiled The missing-tile problem never really went away. See, e.g., James Oberg, "Shuttle Tile Repair Kit,"
New Scientist Magazine, November 15, 2003, http://www.jamesoberg.com
/11152003repairkit_col.html.

New Era Ushered In *Chicago Tribune*, April 15, 1981, p. 1.

The quotes from Ed Whitsett, Joe Allen, and NASA about the MMU all come from
"More Favored Than the Birds: The Manned Maneuvering Unit in Space," by Anne
Milbrooke, in *From Engineering Science to Big Science* (Pamela E. Mack, Ed.), NADA
Office of Policy and Plans, NASA History Office, Washington, D.C. (1998).

The crew of Mission STS-11 NASA's space shuttle–numbering system ended up being
more complicated than it needed to be. For simplicity's sake, the missions were as
follows: STS-1 through STS-9 (obviously, the first nine missions, from April 1981
through November 1983); STS-41B through STS-51L (the next fifteen missions,
from February 1984 through January 1986, concluding with STS-51L, in which
the *Challenger* was destroyed); and STS-26 through STS-135 (the subsequent 110
missions, from September 1988 through July 2011). "STS" stands for "Space Transportation System," a carryover from a 1969 governmental study that recommended a
far more ambitious array of machines. For a discussion of the cumbersome sequencing
designations used between STS-9 and STS-26, see https://www.nasa.gov/feature
/behind-the-space-shuttle-mission-numbering-system.

ASTRONAUT TO FACE SPACE ALONE Jim Barlow, "Shuttle Astronaut to
Face Space Alone," *Houston Chronicle*, February 7, 1984, p.1.

14. THE ACCIDENTAL ICON

almost grandfatherly Richard Leiby and Robert Ebisch, "Backpacking in the Cosmos,"
Sunshine Magazine, February 5, 1984, p. 22.

a mighty push Margo Hamilton, "Reaching for the Stars," *Your Mountain Connection*,
April 2019, p. 18.

A buildup of ice on *Challenger*'s port side Jennifer Ross-Nazzal, NASA Johnson Space
Center Oral History Project, Interview of Robert L. "Hoot" Gibson, January 22, 2016,
pp. 45–46.

That's one small step for Neil I wasn't a fan of the statement at first. The words didn't seem to fit the gravity—microgravity, maybe—of the moment. I wasn't the only one. A *San Francisco Chronicle* writer published a piece in which he bemoaned the informality of the quote, and stated, "The ear flinches." But context is important here. Like pilot humor, astronaut humor is deadpan and self-deprecating. It's Steve Hawley commenting, when shuttle mission STS-41D was scrubbed shortly before liftoff due to a fire on the launchpad, "I thought we'd be higher when the engines quit." Or Michael Collins telling Houston, "If I stop breathing, you'll be the first to know." Donn Eisele, navigator on Apollo 7, was known to complain, "I'm the navigator, and I think I have the right to know where we are." Clearly Bruce McCandless wanted to associate himself with the words Armstrong spoke in 1969, which are among the most memorable in American history. But it's a difficult sentence to match. My dad said he made the comment to lighten the atmosphere back at Mission Control. My mother, who was there, confirmed it had the intended effect. There was a burst of laughter in the room. So maybe it wasn't poetry, but it was respectful, and, first and foremost, practical. It's grown on me over the years. For the *Chronicle*'s assessment, see Charles Burress, "Formality in Space," *San Francisco Chronicle*, February 8, 1984. For a little added context, it's worth noting that the diminutive Pete Conrad said something quite similar when he first stepped down from the lunar module to the surface of the moon during Apollo 12. "Whoopie! Man, that may have been a small one for Neil, but that's a long one for me."

It IS Florida! STS-41B Air/Ground Transcript, 038:13:25, 2/7/84, https://archive.org /stream/STS-41B/Sts-41bA-gTranscript_djvu.txt.

We certainly had a good time https://archive.org/stream/STS-41B/Sts-41bA -gTranscript_djvu.txt.

President Reagan was on the line https://archive.org/stream/STS-41B/Sts-41bA -gTranscript_djvu.txt.

The machine "shuttered, rattled, and shaked" Andrew Chaikin, "Untethered," *Air & Space Magazine*, October 2014, https://www.airspacemag.com/space/untethered -180952792/.

It is a true statement "Meeting Captain Bruce McCandless II," The Rogue Astronaut Blog, https://therogueastronaut.com/2017/11/13/meeting-capt-bruce -mccandless-ii/ November 13, 2017.

One of the great mind-blowers of the 20th century Chaikin, "Untethered," https://www.airspacemag.com/space/untethered-180952792/.

"there is no limit to human potential" Official Statement of Senator John S. McCain, December 22, 2017. McCain also said: "I'm deeply saddened today by the passing of Bruce McCandless, a brilliant aviator and astronaut who dedicated his life to serving the country he loved. Bruce and I were both members of the Class of 1958 at the United States Naval Academy. As an undistinguished graduate of that class, I always looked up to Bruce—not only for his incredible intellect, but also for his character and integrity, which embodied the highest values of the United States Navy."

How did it feel? he was asked Draft, with notes, of French radio interview responses, c. June 1984. McCandless Family Papers.

Back in 1966 Chaikin, "Untethered," https://www.airspacemag.com/space/untethered -180952792/.

When he maneuvered away from his crewmates 178 miles is an approximation. Various sources say *Challenger* was flying anywhere from 165 miles to "330 kilo-meters" (202 miles) above Earth's surface that day. One STS-41B crewmember has suggested to the author that 178 is a reasonable estimate. A spacecraft cir-cling our planet must travel through space (by one definition, "space" is any point higher than fifty miles above ground) at a speed great enough so that it won't fall to the surface but not so fast that the craft won't fly off at a tangent to the globe. This required speed is called orbital velocity. The higher the orbit, the less speed is needed to maintain it. The figure 17,500 miles per hour is a sort of journalistic shorthand, often recycled, to describe the speed of objects in low Earth orbit. In actuality, the number varies from mission to mission depending on the altitude of the spacecraft. As a NASA website put it a few years ago, "like any other object in low Earth orbit, a Space Shuttle must reach speeds of about 17,500 miles per hour (28,000 kilometers per hour) to remain in orbit. The exact speed depends on the Space Shuttle's orbital altitude, which normally ranges from 190 miles to 330 miles (304 kilometers to 528 kilometers) above sea level, depending on its mission." STS-41B flew at an average altitude of around 180 miles above Earth. STS-31, by contrast, flew higher than any shuttle mission before it, roughly 380 miles above the planet. Because it was so much higher, the shuttle's orbital velocity during STS-31 was somewhat lower. And because the orbiter was traveling slower, and was

farther away from the planet, Earth seemed to go by for McCandless on STS-31
at a more leisurely pace than it had during STS-41B. See https://www.nasa.gov
/centers/kennedy/about/information/shuttle_faq.html.

still an EVA record According to NASA, the record-setting flight was the very first
one. It occurred on February 7, 1984, took one hour and twenty-two minutes, and
achieved an operating range (i.e., distance from the orbiter) of 96.93 meters, or
317.9 feet. All in all, five astronauts operated the MMU (or MMUs, since there
were two units) on three different missions in 1984 for a total of ten hours and
twenty-two minutes. Two cosmonauts, Aleksander Serebrov and A. S. Victorenko,
performed EVAs using the Soviet version of the jetpack, variously referred to as
the SPK (for *Sredstvo Peredvizheniy Kosmonavtov*, or cosmonaut mobility equipment) or "space motorcycle," on February 1 and 5, 1990, respectively. The SPK
looked vaguely like the MMU but was larger and considerably heavier, weighing
in at 481 pounds, although, since the SPK was "built in" to the Soviet pressure
suit, this figure may represent the combined weight of the unit and the suit. Just as
with the *Buran*, the Soviets had clearly been paying attention to American technological developments; my father said in a 2004 speech to the Denver Museum
of Natural Sciences that when he examined a Soviet blueprint for the SPK, he
found areas where English language notes related to the MMU had simply been
copied and translated into Russian. According to an Air Force Foreign Technology Bulletin issued in 1990, the unit operated on compressed oxygen rather than
nitrogen and had 32 thrusters, in two redundant sets of 16. It was sheathed in a
soft, removable white material for passive thermal protection as opposed to the
MMU's hard outer surface. The cosmonauts' maneuvers, which were performed
while tethered to the *Mir* space station, totaled one hour and thirty-three minutes.
The Soviets weren't any less daring than their American counterparts. It was just
that, if an SPK unit failed, the station couldn't have retrieved the stranded (or possibly rapidly receding) astronaut, as the shuttle could. Such a rescue by the shuttle
was never performed, of course, but it was seen as a viable option. As MMU project
manager Bill Bollendonk put it to a worried Martin Marietta board of directors on
the eve of the launch of STS-41B, in February 1984: "Imagine a Ferrari going after
a Volkswagen. That would be the orbiter going after an astronaut in a malfunctioning MMU. We could definitely catch him." MMU and SPK statistics, James
W. McBarron, Charles E. Whitsett, Guy I. Severin, and I. P. Abramov, "Final

Review Draft," "Individual Systems for Astronaut Life Support and Extravehicular Activity," for *Foundations of Space Biology and Medicine*, September 1991; Lt. Christine Helmke, "Soviet Manned Maneuvering Unit," FTD-2660P-127/58-90, March 26, 1990; and Bill Bollendonk story, Interview of Bill Bollendonk, Conifer, Colorado, August 13, 2020.

performed as planned I was tempted to say "performed perfectly," but Gardner did report one problem that arose during his use of the MMU on STS-51A: sluggish acceleration and high thruster activity (and resultant high propellant usage) in certain translation modes. It's unclear whether the problem was remedied, as the MMU was never flown afterward. Bruce McCandless and his colleagues at Martin Marietta continued to envision upgrades for the MMU well into the nineties. Among such improvements were digital control electronics, data capture and recording/transmission capability, potential use of space station waste gases for MMU propulsion (!), and the development of a GPD receiver and location display for the MMU operator, with such technology presumably to take the place of the solid-state range finder McCandless used in 1984 to gauge the distance he'd traveled from *Challenger*. As a side note, following Gardner and Allen's rescue of the satellites deployed by STS-41B, Palapa B2 and Westar VI were both repaired on Earth and returned to space via rocket.

The timing almost matches up Michael Cassutt, "Secret Space Shuttles," *Air & Space Magazine*, August 2009, https://www.airspacemag.com/space/secret-space-shuttles -35318554/. STS-27 holds another distinction as well. The mission's orbiter, *Atlantis*, came close to disintegrating upon reentry due to tile damage caused by pieces of foam insulation striking the orbiter on launch. Crewmembers were able to see evidence of the damage during the mission and knew it was quite possible that they would die. Luckily, the tile damage occurred on a noncritical area of the orbiter, and the crew survived. Fifteen years later, the crew of *Columbia* was less fortunate. A similar accident led to destruction of the spacecraft and the death of all aboard during reentry in 2003.

NASA's 1993 SAFER training manual Scott Bleisath, *Simplified Aid for EVA Rescue (SAFER) Training Manual*, NASA Lyndon B. Johnson Space Center Publication, JSC-26283, July 23, 1993, p. 7.

Weighing in at 85 pounds, SAFER fits are worn by Russian cosmonauts as well. The device exists in two versions—one that works with the American EMU and another designed to be used with the Russian ORLAN pressure suit. Direct descendants of the MMU, SAFER units protect crew members of the International Space Station on every space walk.

the perfect way to maneuver around the irregular surfaces of asteroids NASA canceled its contract with Martin Marietta for maintenance of the MMU in 1988. The company implored the agency to reconsider. In a letter to JSC Acting Director Aaron Cohen, one company executive wrote that it would be willing to front the money needed for development of a prototype Central Electronics Unit for the MMU and expressed Martin Marietta's hope to "prevent this highly successful and vital national resource from becoming 'moth-balled' and perhaps never flown again." In March 1991, in a personal letter to Cohen, Bruce McCandless noted that while he was now working as acting chief engineer for Martin Marietta Astronautics Group in Denver, he still had "a special interest in the Manned Maneuvering Unit," and that his hope was "to see it flown again in the near future, possibly carrying the IMAX camera for shots looking back at an Orbiter passing across the Earth's surface. Such shots would be truly magnificent."

In April, McCandless and his colleagues authored a presentation they called "MMU Return to Flight." Regarding possible uses of the MMU, the brief lists crew rescues and the ability of a MMU-equipped astronaut to retrieve satellites and to move large objects, as might be needed for construction of a space station. Under the heading *A Potential Problem Without a Demonstrated Solution* they listed, "If Disabled Spacecraft Has Significant Tumble Rate, MMU May Be the Only Option." Later that year Bruce McCandless met with a representative of the National Air and Space Museum to discuss the possibility of the museum paying for MMU time aboard a future shuttle flight to make another IMAX movie along the lines of *The Dream is Alive* and *The Blue Planet*. Martin Marietta was still attempting to convince NASA and/or private industry to use the MMU as late as 1992. At a meeting with NASA representatives in April of that year, for example, the company listed "numerous possible applications of the MMU," including using the machine to study and characterize shuttle glow emissions in the UV and IR spectrum, obtaining imagery of wastewater dumps to provide data for Space Station Freedom design, and using the MMU as a space-borne forklift or cherry picker for use in building the space station. Meeting attendees also discussed NASA's renewed interest in "thermal tile inspection and repair capability"—ironically, one of the tasks the MMU was developed to address some twenty years earlier but never actually did. The Draper Company of Boston has developed devices it says could take the place of the MMU's lineal descendant, SAFER. Their machines would be able to automatically return an astronaut to his or her spacecraft in the event of the operator's unconsciousness. The company also claims its specialized suits "could enable astronauts to explore asteroids and repair satellites, as well as make it easier for them to live and work in and around space stations," which sounds like it is singing from the same hymnbook as Lockheed Martin. Asteroids contain minerals and other materials of interest to scientists, planetary theorists, and venture capitalists. In 2015, President Barack Obama proposed a crewed mission to an asteroid by 2025. Interestingly, Bruce McCandless claimed in an interview in 2015—the same year—that NASA was working on "an advanced version of the MMU for the exploration of asteroids." Unfortunately, asteroid visits are not currently an American space priority, and it is unclear how much progress, if any, was made on improvements to the MMU. For the interview, see Hannah Booth, "That's me in the picture: Bruce McCandless, 47, in the world's first untethered space flight, February 1984," *The Guardian*, July 10, 2015, https://www.theguardian.com/science/2015/jul/10/bruce-mccandless-first-untethered-space-walk-challenger.

a beautiful example of aerospace engineering Andrew Chaikin, "The Story of NASA's Jet-Propulsion Backpack," *Smithsonian Magazine*, April 2014, https://www.smithsonianmag.com/science-nature/story-nasas-jet-propulsion-backpack-180950190/.

The only way you could make it easier *Ibid.*

15. INTERMISSION

the graying Buck Rogers was on his way out Robert Reinhold, "Interviews Disclose Old, Sometimes Bitter Resentments in Astronaut Corps," *The New York Times*, April 3, 1986, Section 3, p. 3

Early 1986, boom Kevin Rusnak, NASA Johnson Space Center Oral History Project,

Interview of Dr. Joseph P. Kerwin, May 12, 2000, https://historycollection.jsc.nasa
.gov/JSCHistoryPortal/history/oral_histories/KerwinJP/KerwinJP_5-12-00.htm.

hoped for a front-seat flight role Bruce McCandless II, Letter to Lt. Gen. James J.
Abrahamson, Associate Administrator for Space Flight, NASA, October 31, 1983.
McCandless Family Collection.

16. LAUNCHING THE TIME MACHINE

a chaotic soup of gas NASA online publication, "The Hubble Story," retrieved October
7, 2020, https://www.nasa.gov/mission_pages/hubble/story/the_story.html.

two snowmobile suits Charlie Wood, "Kathryn Sullivan: Spacewalker and Earth
Explorer," *Space.com*, January 13, 2020, https://www.space.com/kathryn-sullivan
-bio.html.

a top-to-bottom inspection of Hubble Christopher Gainor, *Not Yet Imagined: A Study
of Hubble Space Telescope Operations* (Washington, D.C.: National Aeronautics and
Space Administration, Office of Communications, NASA History Division, 2020),
p. 90.

developed a tool-tethering system See, generally, "Another Dividend from Air and
Space," in *NASA Activities*, Vol. 14, No. 5, May 1983, p. 28.

Senator Barbara Mikulski of Maryland David L. Chandler, "Telescope and Shuttle
Problems Making It a Hard Year at NASA," *Miami Herald*, July 23, 1990, p. 20.
See also "Techno-Turkey Needs Quick Fix," *Springfield (MO) News-Leader*, July 2,
1990, p. 4.

17. EYE IN THE SKY

There were weird ideas Leonard Davis, "Hubble at 20: Reflections on the Universe,"
Aerospace America, April 2010, p. 30.

Much of the light caught by the Deep Field Ross Andersen, "Golden Eye," *Los
Angeles Review of Books*, February 15, 2012. https://lareviewofbooks.org/article
/golden-eye/.

a science celebrity David Devorkin and Robert W. Smith, *Hubble: Imaging Space and Time*, (Washington, D.C.: National Geographic, 2008), p. 7.

Akin to a modern version of a torch-wielding mob *Ibid.*, p. 6.

As astronomer Abraham Loeb puts it Abraham Loeb, "Can the Universe Provide Us with the Meaning of Life?" *Scientific American*, January 21, 2020, https://blogs.scientificamerican.com/observations/can-the-universe-provide-us-with-the-meaning-of-life/.

Bruce played a key role in helping the team James H. Crocker, Letter to Admiral Robert Natter, USN (Ret.), Chairman, Distinguished Graduate Awards Committee, July 10, 2006, p. 1.

18. A NEW START

chart Source: https://ntrs.nasa.gov/search.jsp?R=19920074637 and email correspondence with Dr. Ben Clark dated October 12, 2020, and November 16–18, 2020.

He published a long essay Bruce McCandless II, "Revere, Then Retire The Shuttle," *The Denver Post*, February 23, 2003, p. E-1. My father's op-ed about the Nowak case, "NASA Acted Hastily," was published in *The Denver Post* on March 9, 2007. Dad's posthumous paean to the first man on the moon, "An Astronaut's Tribute to Neil Armstrong," was also published by the *Post*, on August 30, 2012. "Thanks, Neil," he concluded on that occasion. "We will never forget you." Until my mother died, a year and a half later, these were the most emotional words I'd ever heard him utter—or *write*, anyway. My father was not alone in his admiration for Armstrong, of course. Among space enthusiasts, only a few individuals are customarily referred to by their first names: Deke. Gus. Buzz. Maybe Wally. But only one name is said like a blessing: *Neil*. And when we hear it, most of us do.

ABOUT THE AUTHOR

B ruce McCandless III grew up in El Lago, Texas during the Apollo and Skylab eras. He graduated from the Plan II Honors Program of the University of Texas in 1983 and went on to earn degrees from the University of Reading in England and the University of Texas School of Law. After teaching at Saint David's School in New York City, he returned to Austin to practice law. He is the author of *Sour Lake, Beatrice and the Basilisk,* and, with his daughter Carson, *Carson Clare's Trail Guide to Avoiding Death.* Bruce serves on the board of directors of the Worthy Garden Club, an Oregon-based environmental organization, and the Austin Public Library Foundation. He and his wife, Pati McCandless, live in Austin.